MAN AND THE WINDS

MAN
AND THE WINDS

E. AUBERT DE LA RUE

Translated by
MADGE E. THOMPSON

PHILOSOPHICAL LIBRARY : NEW YORK

Printed in Great Britain by
WILLIAM BRENDON AND SON LTD
THE MAYFLOWER PRESS
(late of Plymouth)
WATFORD
HERTS

CONTENTS

LIST OF ILLUSTRATIONS

7

TRANSLATOR'S NOTE

IN translating this book I have omitted some passages of purely local interest in France but, on the other hand, where the text has suggested them to me, I have added a few relevant notes in the hope that they will be of interest.

Since the book first appeared some of the subjects dealt with have assumed greater importance. This is particularly the case in the field of wind power, which is now being seriously studied on an international scale and in which notable developments have taken place in the last few years. Considerable progress has also been made in studies of the effects of the wind in aerial navigation, and the part played by the wind in soil erosion is receiving greater attention in view of the present-day interest in the development of the arid or semi-arid zones of the world for food production.

The Centigrade scale of temperatures has been retained in the translation, but for the benefit of those who may be more familiar with the Fahrenheit scale a conversion table is given overleaf. In converting values in the metric system into English units the figures have been rounded off where they are obviously only approximations.

February, 1955   M.E.T.

Centigrade – Fahrenheit Equivalents

°C.	°F.	
100	212	Boiling point (*of water*)
90	194	
80	176	
70	158	
60	140	
50	122	
40	104	
35	95	
30	86	
25	77	
20	68	
15	59	
10	50	
0	32	Freezing point (*of water*)

AUTHOR'S PREFACE

THE study of the wind comes within the province of meteorology, but its effects directly concern man and the aim of this book is to show how he behaves in the presence of this element. One has only to live in a rather windy climate to realize how important the movement of the air is and how varied, and often unexpected, are its consequences. The presence on roof-tops of wind motors—not to call them windmills, which are already things of the past in most parts of the world—the hedges which protect houses and the windbreaks of trees sheltering crops and even railways, are the most characteristic and obvious signs of the frequently agitated state of the atmosphere. There are many other signs testifying to the constant care of the inhabitants to mitigate the attacks of squalls. Do those who enjoy a pleasantly soft breeze in summer suspect how much the wind may be dreaded in other places and how much its perpetual blowing eventually becomes an obsession—not to mention the damage caused by its violence?

Enormous areas on our planet are almost entirely dominated by the wind. In the solitudes of the polar regions and the deserts it is, so to speak, the only element of life. The traveller who ventures into these areas has a continual struggle against the elements. Atmospheric currents fortunately do not rage everywhere with the same intensity, but in every country examples could be found of real "wind corridors", deservedly notorious for the violence of the air currents which blow through them and for the havoc which these cause. Among the most favoured regions, enjoying a relatively calm climate, none can boast of being entirely free from accidental disturbances giving rise to gales. These change their name from one country to another, but they almost always have the same disastrous results.

In the presence of a natural force so formidable and so capricious, the attitude of man is twofold. He seeks ways of using it, but he tries, above all, to defend himself against it. Mention of the use of the wind quite naturally calls to mind the

11

sail—for centuries the only means of propelling ships—and the picturesque windmills of former days. These were the two great uses of the wind, but there are others, less widely known and more modest. Even now, when we have almost ceased to call on this source of energy, it is still of service.

It happens that my travels round the world have given me the opportunity to become familiar with most types of wind, from the sand storms of the desert to the tropical cyclones, from the trade winds to the great west winds of the South Seas. I have been able to see in many countries the way in which people make use of the wind and protect themselves from it. Many of the examples given in this book, therefore, come from personal observation.

There is no lack of books dealing with the wind or devoting large sections to it, and they have been of great help to me. It does not appear, however, that the subject of the wind, considered from the point of view of Human Geography, has yet been treated in a detailed manner. In writing this book, I have tried to show how complex are the relations between man and this formidable element.

Aubert de la Rüe.

CHAPTER I

Types and Names of Winds

THE wind is only air in motion due to the variations of density produced, at different points of the atmospheric mass, by the action of the sun's heat which is unequally distributed over the surface of the globe. The air is displaced from the centres of high pressure, which are concentrations of heavy air, towards the centres of low pressure, corresponding to concentrations of light air.

Meteorology

The general circulation of the atmosphere is determined by the existence round the earth of four zones of constant high pressure, which are separated by three belts of permanent low pressure. There are the two polar areas of very high pressure and the two sub-tropical areas of high pressure lying between the latitudes of thirty degrees and forty degrees in both hemispheres. The largest belt of low pressure lies on each side of the equator, while two smaller belts are just below the Arctic and the Antarctic Circles. On the whole, the low-pressure zones correspond to areas of relative calm.

The importance of these successive belts of high and low pressure is due to the existence of atmospheric currents between them. At the two poles, above the ice caps of Greenland and the Antarctic, there is a régime of divergent winds moving from the interior towards the coast. Beyond that, up to the limit of the polar circles, the prevailing winds come from the east. Between the latitudes of sixty degrees and forty degrees the great west winds, which characterize the temperate zones, meet.

13

Their direction is south-west in our hemisphere and north-west in the southern hemisphere. This is also the domain of the great cyclonic depressions, travelling from west to east and causing the strong Atlantic gales and the formidable storms of the South Seas. Approaching the equator, we find a relatively calm zone of variable winds, which corresponds to the sub-tropical belts of high pressure. Towards the latitudes of thirty degrees north and thirty degrees south the *trade winds* begin, blowing respectively from the north-east and from the south-east. Separated by the band of equatorial calms, the trade winds of the two hemispheres are the most regular and extensive winds in the world. They have the same sphere of influence as those violent tropical storms known as *cyclones* and *typhoons*. Such is, in its broad outlines, the general wind régime on the earth's surface. We shall see that there is also a host of other winds determined by geographical conditions.

Man could not remain indifferent to this enormous agitation of the atmospheric mass, which presents very diverse characteristics. His attitude to the wind is twofold: to use it when he can and to defend himself from it. In this respect, winds cannot be divided into "good winds" and "bad winds" according to whether they appear useful or inimical to man. The same wind is sometimes beneficial, sometimes harmful, depending on the time and place where it is felt, on its intensity and on the local characteristics which it may acquire.

Among the essential characteristics of the wind are its direction, its force, its degree of heat or cold, of dryness or humidity, the physical effects of its action and the regularity or irregularity of its régime.

The direction of the wind is always expressed by the quarter from which it blows. It rarely, however, maintains a constant direction for a definite time; generally it oscillates about an average direction, a fact which is clearly shown by the haphazard movements of a weathercock. Its speed is also a very variable factor. The greater the difference of pressure between the two points between which it is blowing, the greater is its speed. Moreover, the movement of the air is strongly influenced by the natural obstacles that it encounters. The wind, which is composed of a bundle of practically parallel air streams, is generally affected by the principal prominences and irregularities on the earth's surface, and this causes it to undergo enormous variations in force and in direction. That is why, near the ground, it never blows uniformly, but often in violent gusts which rise and die down in a few seconds. L. Cattala (Ref. 17) mentions a storm in the Isle of Ré, where the wind speed rose from twenty-seven to forty-seven miles per hour in the space of one-third of a second. Even when it seems constant, very delicate instruments show that it changes considerably in very short intervals of time. The cause of these irregularities is not always apparent. Nearly everyone has noticed on very windy days how grass bends and rises again in undulating and uneven lines, showing the infinite variations of the breeze. Strong trees are sometimes uprooted or broken in a storm while, a short distance away, others, more fragile, remain unharmed. On the sea, where nothing impedes the movement of the wind, the uneven gusts which here and there disturb the surface of the water are still more noticeable.

Wind Speeds

Wind speed is expressed in miles per hour or in metres per second and a graded system has been established—the Beaufort Scale—which divides the winds into twelve categories according to their speed. In Europe, the maximum wind speed seldom exceeds ninety miles per hour, even in a gale near the coast, but at exposed sites near the west coast of Britain gusts of around 100 m.p.h. have been measured during the winter. In an exceptionally strong gale in January, 1953, gusts of 125 m.p.h. were measured in the Orkney Islands and the average wind speed was about 90 m.p.h.—*Trans.*

In tropical countries subject to cyclones the wind is more violent than it is in Europe. Cattala quotes average maximum wind speeds for the east coast of Madagascar of the order of 135 and 180 miles per hour in very strong gusts, remarking at the same time that measurements have not yet been made *near* the centre of these cyclones, where the speed of the wind is certainly greater still. He also says that Oliver Z. Fassig, a meteorologist in the United States Weather Bureau, estimates that the momentary speed of gusts near the centre of a typhoon is nearly 200 miles per hour.

[It will be noted that the reference is to *near* the centre; *at* the centre (in "the eye of the storm") the wind speed is nil (see page 48).—*Trans.*]

No part of the world is entirely free from wind, but there are places with much more wind than others. Some countries are well known for their frequent and violent winds, others, more fortunate, enjoy a much calmer atmosphere. It is probably true to say that the only places where wind is really rare are the large dense forests of the equatorial belt and a few very sheltered valleys in the midst of groups of high mountains.

On fine days the calm is often deceptive. In fact, in a scarcely perceptible breeze, the contour of the ground may give rise to vertical eddies. These convective movements are called turbulence.

Kinds of Winds

There are many types of winds, distinguished from one another by certain well-defined characteristics. First of all, they can be divided into two main groups: the *regular* or *régime winds* and the *accidental winds*. The first comprises, on the one hand, the *constant* or *prevailing winds*, such as the *trade winds* in the tropical zones and the great *west winds* of medium latitudes, and, on the other hand, the *periodical* or *seasonal winds*, like *monsoons* and *solar breezes*, as well as *day* and *night breezes, land* and *sea breezes,* and *mountain* and *valley breezes.* The *accidental winds*, caused by atmospheric turbulence, include the gales in the temperate zones, the cyclones in the tropics and, in general, the winds peculiar to each country, such as the *mistral* in Provence, the *föhn* in the Alps and the *Pampero* in Argentina. Although they are accidental, most of the winds included in this category have a seasonal character. Thus, in Europe, storms are more frequent during the winter months, while in the tropics the cyclones occur in the hot season.

Besides these principal winds, many others are known to meteorologists. There are the *variable breezes*, which blow sometimes from one direction, sometimes from another, without regularity either in force or in direction. To these are related *changeable winds*, of short duration, which occur after a storm and before the normal régime is re-established.

At one time winds were thought to be perfectly horizontal but, although they usually have a prevailing component which is horizontal, they certainly have ascending and descending components as well. In contrast to the *surface winds*, familiarly called *weathercock winds*, there are the winds of the upper atmosphere at high altitudes; these, as can be seen from the movement of the clouds, are usually different in direction and intensity from those which, as they are felt near the ground, are of greater interest to man. A final distinction has to be made —that between *local breezes* affecting only a restricted area and *general winds* related to the great atmospheric currents.

Names of the Winds

The names of the winds have many different origins. The simplest way of designating them, employed by meteorologists and sailors, is by the direction from which they blow. Usually it is sufficient to use the eight principal points of the compass which constitute the wind rose. In the days of sailing ships, when the precise direction of the wind was very important, twenty-four intercalary divisions were added to the main divisions. Each of the thirty-two thus obtained, corresponding to a thirty-second part of the circumference, was called a "point of the compass"—$11\frac{1}{4}°$.

In most countries it is usual, in current speech, to name the winds either according to their geographical origin or according to one or other of their specific characteristics. The names can therefore be deceiving. Sometimes the same wind changes its name in different districts; sometimes, on the other hand, the same name is applied to atmospheric currents of different origins. The case of the *south-west wind*, which brings rain and may be beneficial or harmful according to the season, is an example of this. In the Juras and the Alps, from the Lake of Geneva to Mont Ventoux, it is called the *wind of the Midi* (or south wind), although it does not come from the south and has an Atlantic, and not a Mediterranean, origin. In those districts it is often simply called the *wind*. When people speak of the wind in Geneva it is unnecessary to specify it, everybody

B

knows which it is. To many people, the south-west wind is the *rain-bearing wind,* as P. Deffontaines (Ref. 26) observed in the Moyenne-Garonne, where it is also called the *wind from below* or the *wind from down under,* that is to say, coming from down-stream—expressions very similar to the *wind from down below* which is used in the Alps to denote this same wind and by which is meant that it comes from more southerly regions. I have come across this term in the Province of Quebec on the banks of the St. Lawrence, where, it may be said in passing, the local names of the winds are nearly all inaccurate. There it is applied to the rain-bearing east wind blowing from the Atlantic, while the west wind coming from up-stream is the *wind from above.* The south-west wind in Europe—the sailors' *sou'wester*—is called the *labech* in Marseilles and the *libeccio* in Corsica. It is also very often given a local name from its immediate geographical origin. It is the *Grenoble wind* at Combe-de-Savoie and the *Geneva wind* on the shores of the Lake of Geneva and in the Canton of Vaud at the foot of the Juras.

Among the winds which frequently change their name, we may mention the *Lombarde* of the Maurienne, or *blizzard,* which piles up snow on the mountain roads, making them impassable, and which, as E. Bénévent tells us (Ref. 5) is often given the name of the pass through which it penetrates into France. At Val d'Isère, it is called the *Galise wind,* at Ste.-Foy the *mountain-pass wind,* at Sécz and at Bourg St.-Maurice the *little St. Bernard wind.* The Alps provide another example of a wind which is constantly changing its name; it is the *traverse,* a term which is used to designate the west wind in the foot-hills, while in certain parts of the Alps themselves it often has a particular name in each village. The west wind thus becomes the *wind of Arsine* at Monêtier-les-Bains, the *wind of Montmélian* on the left bank of the Isère, the *wind of Armenaz* on the Albertville side, etc.

In reviewing the names of winds we can find some which certainly do not lack originality. Here are a few examples collected by G. Marcy in the course of a study (Ref. 49) of the Moroccan names of the local winds. There are about fifty in number, Arab as well as Berber. The west wind, in Arabic *gharbi,* is often called *laâwan* (helper) because it helps in winnowing the grain in the threshing season. A particularly violent *sirocco* which occurs in certain exceptional years is called *arifi* at Bou Debib and Safi, *irifi* at Taza and *rifi* at Rabat and in the Chaouïa. They are all forms of the Berber word *arifi*, derived from *aref* and signifying "to heat on the embers, to scorch". *Mezzer-ifoullousen* (that which plucks the fowls) is a picturesque Berber expression to designate a cold and violent south-east wind.

Weather Forecasting

For centuries men have looked at the sky and studied the moon, the wind and the clouds, in the hope of discovering what the weather will be. The empirical rules formulated from these observations have given rise to a host of popular sayings in the country, some of them reliable, others much less so.

The barometer never shows us more than the existing state of the atmosphere and the indications given by the wind are essential for forecasting the weather. By observing whence it blows and examining the appearance of the sky, especially the formation and movement of the clouds, we can predict, with some degree of certainty, the immediate change in the direction of the wind and deduce from it the change in the weather which it will bring. In fact, the rotation of the wind normally takes place in the same direction in each hemisphere. In ours its gyration is normal when it passes from the west to the east by the north. When, by chance, it passes from the west to the east by the south, sailors say that it turns against the sun and rightly mistrust these *backward-moving winds*. An English proverb neatly expresses this fear:

> When the wind veers against the sun
> Trust it not, for back it will run.

This observation is not new and there appears to be a reference to it in the Bible (Ecclesiastes i, 6): "The wind goeth toward the south, and turneth about unto the north; it whirleth about continually, and the wind returneth again according to his circuits."

Weather forecasting, which often means, in fact, predicting

the probable wind, as variations of temperature and humidity largely depend on it, has acquired particular importance since the development of aviation. It is, however, a very old problem, which has always been of great interest for navigation, especially in the days of sailing ships, and for agriculture. Sailors are much concerned with the force and direction of the wind and with the fact that certain winds bring fog and others dissipate it. To the latter, the fishermen on the Banks of Newfoundland give the name of *fog-eaters*.

In 1860, a valuable service to shipping was started in France at the instigation of Le Verrier, who proposed to the French Admiralty that the existing meteorological network should be used to follow the track of storms over Europe and warn the ports of the approach of gales. A similar system was organized shortly afterwards in England. Since then, all the coastal regions have possessed meteorological stations in ever-increasing numbers, sending out weather bulletins and gale warnings by radio to state ships and aeroplanes. Some of the stations, like Stornoway in the Hebrides and Valentia in Ireland, are particularly important for announcing to western Europe the approach of depressions from the west.

From the meteorological point of view the needs of agriculture are different from those of navigation. The effects of the wind concern the farmer more than its frequency. He is interested in its force, in the changes of temperature that it causes, in the frost that it may bring to endanger his crops, and in the rain or fine weather that he can expect from it.

The weather forecast is a matter of interest to nearly everybody, even to town-dwellers, although to them it may be only a guide in arranging their outdoor amusements.

CHAPTER II

Règime Winds and Accidental Winds

THE many winds which blow over the earth's surface can actually be reduced to a fairly small number of types, some of which deserve special attention. Their effects on human activity are both profound and varied and depend on whether they appear as a useful force or as a danger.

The Trade Winds

Although the existence of constant winds round the globe on either side of the Equator and the useful part that they could play in navigation had been revealed by the great maritime expeditions of the fifteenth century, the origin of the *trade winds* remained for a long time obscure. Buffon (Ref. 14), without adopting it himself, quotes the extraordinary explanation for them given at one time by Lister. "Dr. Lister, who was in other respects a good scientific observer," he wrote, "claims that the general east wind blowing in the tropics all the year round is produced solely by the respiration of the plant called sargasso (*Fucus natans*), which is very abundant in those latitudes, and that the different winds over the earth's surface are caused only by the different distribution of trees and forests. He solemnly gives a ridiculous explanation of the origin of the east wind, saying that at midday the wind is stronger because the plants are hotter and breathe more often and that it blows from east to west because they behave rather like sunflowers and always breathe on the side facing the sun."

The first serious theory to explain the regular phenomenon of the trade winds was that of the English astronomer Hadley

(1735). It is based on the existence of north-south currents caused by the differences of temperature between the polar and equatorial regions. Over the hottest parts of the oceans the over-heated air becomes lighter and lighter and rises in the atmosphere. This is the region of the equatorial calms, where the air is practically motionless and the sky always overcast and where torrential rain is often accompanied by thunderstorms. The rising air is continuously replaced by new masses of air which, sliding over the surface of the ocean, rush in from the north and the south, but the rotation of the earth tends to make them turn more and more towards the west as they approach the torrid zone. The belt of calms which separates the trade winds of the north-east from those of the south-east has no definite boundary. It extends slightly northward in our summer and southward in the winter, but never encroaches much on the southern hemisphere.

The trade winds are not really as constant as they are generally supposed to be and their régime includes periods of relative calm. Everywhere in the tropical zone it can be observed that they are specially strong and well established for a period of six months, corresponding to the cold, dry season, that is, from November to April in the northern hemisphere and from May to October in the southern hemisphere. They are not so strong and regular in the hottest months of the year, which usually correspond to the rainy season. Then they are frequently interrupted by periods of calm.

The Monsoons

The China Seas and the Indian Ocean are the centre of a remarkably extensive seasonal phenomenon which constitutes an exception to the general system of atmospheric circulation. It is a perfectly regular wind régime called the *monsoons* (from the Arabic *mausin*, meaning season). There are two monsoons and they divide the year into two equal parts. The *north-east monsoon*, which mingles with the *trade wind* from the same direction, begins in November and ends in March. After a few weeks of calm and of light variable breezes, the *south-west monsoon* begins in May and finishes in October.

The *north-east monsoon* is caused by the difference of temperature between the hot zones near the equator and the regions of Central Asia, which are cold in the winter from the month of November onwards. It is not a very strong wind and chiefly occurs during fine weather in the Indian Ocean. The sky is clear and the air relatively dry.

The *south-west monsoon*, the real *monsoon*, is quite different. "The lowering of the sub-tropical high pressures over Asia in summer," observes E. de Martonne (Ref. 51), "continues until they are low enough to draw in the air from all over the Indian Ocean, attracting the southern trade wind in the northern hemisphere as far as the tropics." The north-east monsoon then gives way to the south-west monsoon and air is sucked from the equatorial regions, which are cooled at this season of the year by heavy rains, towards the great plateaux of Central Asia, the Arabian Sea and the Persian Gulf. The greater the difference of temperature, the stronger this suction is, hence the violent character of the south-west monsoon. From the end of May the weather changes abruptly; the sky becomes over-cast, the moisture-laden air is hot and heavy and the sea is swollen by the first gales coming from the south-west. At the end of July and during the month of August the monsoon attains its greatest force. The wind dies down at the end of September, the sea becomes calm and the sky clears.

The south-west monsoon is sometimes so strong that ships returning to Europe from the Far East are obliged to abandon the normal route to avoid the rough seas. Instead of passing by Minicoy, they may go as far south as the Channel of Cardiva to find a calmer zone; some even take the Channel of "One Degree and a Half".

Etesian Winds

If the word is taken in its broadest sense, monsoons can be found in nearly every region of the globe. In North Africa, for instance, the currents coming from the Atlantic and bringing the seasonal rains are often spoken of as the *west monsoon*.

In ancient times the existence of true monsoons was recognized in the eastern Mediterranean and they were widely used for navigation. The seasonal winds from June to September, blowing from north to south over the Ægean Sea, were given the name of *etesian winds* (from *etos*, season). Greek sailors today call them *meltemi*, and, although for a few weeks they are of a rather stormy character, their regularity mitigates their danger. In the western Mediterranean, too, the prevailing winds, caused by the suction of hot air over the Sahara, come from the north. The stunted vegetation of the northern slopes of Majorca and Minorca and the trees bent towards the south plainly show how frequent these winds are.

Local Winds

A distinction has to be made between local winds. Some are local only in their name, which may vary in different places; usually they can be identified with one or another of the great atmospheric currents. E. de Martonne regards as local winds the air currents that are determined by barometric pressure and whose particular characteristics are due to the influence of local geographical conditions. In many districts they have been given special names on account of the great force they sometimes attain and the changes in temperature and humidity they bring. Most of these winds, however, belong to a small number of types, such as solar breezes, or they are connected with the movement of depressions. This is the case with the *föhn* in the Alps, the *mistral* in Provence, the *tramontana* in Roussillon and the *sirocco* in North Africa. They are always more pronounced in the temperate zones than in the tropical.

The monographs devoted to the study of regional climates, of which the very important work of E. Bénévent (*loc. cit.*) on the climate of the French Alps is an excellent example, show the number and diversity of local breezes occasioned by the relief, the orientation of chains of mountains and the difference in temperature between two neighbouring points. One volume would certainly not suffice to enumerate all the winds which blow over the surface of the earth and which, in any case, could not be catalogued. Many, of course, affect only a very limited area—sometimes simply part of a valley. Often, however, these local breezes are of much more direct concern to the mountain-dweller, the countryman, the fisherman and the sailor than the general winds. An accurate knowledge of them is particularly important to the farmer; it is in the country, therefore, that the largest number of names are found to designate winds according to their direction, the season at which they blow and their various characteristics, such as their temperature, their relative humidity, their force, whether they bring rain or fine weather and whether they are favourable or unfavourable to crops.

Land and Sea Breezes

On the coast when the weather is fine, the breeze blows in the daytime from the sea towards the land, which is more quickly heated by the sun. This sea breeze may be felt as far inland as twenty or twenty-five miles. An opposite current is established at night, from the land, cooled by nocturnal radiation, towards the sea whose temperature has hardly changed. The sea breeze starts at eight or nine o'clock in the morning and the greater the heat, the stronger it is. It increases steadily until it reaches its maximum towards two or three o'clock in the afternoon. At sunset there is a period of calm until the land breeze begins. This is generally not so strong as the sea breeze.

The land breeze carries out to sea the layers of air which have been heated up during the day. Along mountainous coasts, however, this nocturnal current of air, coming down from the heights, causes an appreciable drop in temperature. This is the case at Tahiti, where the *hupe*, after sunset, rushes down from the colder mountains to the sea so that, however hot the day has been, the nights are always pleasantly cool and fresh.

All those who live near the coasts in the tropics learn to appreciate the relatively fresh sea breeze which mitigates the heat of the sun and makes the hot hours of the day more bearable. Sometimes, when it is keen, it may become a hindrance to navigation in certain estuaries and lagoons, such as those of the Gulf of Guinea, from the Ivory Coast to Gabon, where, on account of the strong swell, the canoes cannot safely go out in the afternoon.

At Valparaiso, on the coast of Chile, which is subject to the régime of the west winds, the sea breeze (*virazón*) is reputed to be often extremely violent. This is because its effect is added to that of the general winds. Occasionally, it blows there so furiously in the afternoon that the traffic in the town is stopped, the Almandral is deserted and all shipping in the port is at a standstill. On the other hand, the nights are generally calm, for the *terral* or *pulche* (land breeze) blows in exactly the opposite direction to that of the general winds and the two currents partly neutralize each other.

Fisherman often take advantage of these recurring diurnal and nocturnal breezes to put out to sea at dawn and return to port in the afternoon.

Lake Winds

Phenomena analogous to land and sea breezes, when they are not masked by stronger general winds, may also be observed on the edge of great lakes. They are very marked at Chicago on Lake Michigan. People living on the shores of the Lake of Geneva know them well and have given them various local names. The lake breeze coming from the south between ten o'clock in the morning and four o'clock in the afternoon is called the *rebat* on the Swiss side, except at Geneva where it blows from the Great Lake and is called the *séchard*. The nocturnal land breeze, particularly keen between Ouchy and Rolle, is the *morget*, which drops between seven and nine o'clock in the morning. The *morget* mingles with the mountain breeze and is felt as far as the Savoie side. The same phenomenon occurs on the Italian lakes, where the land breeze is at the same time a valley breeze known as *brava del lago* or *ora* (Lake Lugano and Lake Como) or *inferno* (Lake Maggiore), while the nocturnal breeze coming down from the mountains receives the name of *tivano*, sometimes also *tramontana*. In turn, these diurnal and nocturnal breezes push the boats from one shore to the other.

The *montagnère* in Lower Provence and on part of the southern slopes of the Alps is the same as the land breeze. There the diurnal breeze coming from the west is the *séguin* (a Provençal word meaning "following the sun"). In some places, especially in the valley of the Huveaume (Bouches-du-Rhône) the afternoon sea breeze is called the *ponent*.

Mountain and Valley Breezes

In mountainous regions, when the weather is fine, ascending and descending currents blow alternately through enclosed, straight valleys. During the hottest hours of the day the current moves upwards with varying force, which increases with the narrowness of the bottom of the valley and tends to diminish towards the highest point. This valley breeze begins between nine and ten o'clock in the morning. After an interval of calm a movement of air in the opposite direction is set up. This is the mountain breeze, which attains its greatest force at the coldest time in the very early morning.

The general winds sometimes modify the diurnal and nocturnal breezes in the mountains but they are not always strong enough to nullify them completely. If the former blow through a valley towards the top of the mountain, they reinforce the rising current and check that moving in the opposite direction.

During the summer the mountain breezes lower the temperature and make the nights more agreeable. On the other hand, in the winter these descending currents are as cold as the north wind. In valleys where the diurnal breeze is particularly strong, it checks the growth of branches on the side of the tree trunks facing down the valley and gives the trees a permanently bent form. It hinders the traffic, fatigues the traveller on foot and raises clouds of dust. The valley breezes carry up the smoke, dust and water vapour, which collect during the day round the top of the mountains. Shortly after midday the peaks are covered with clouds, which are dispersed by the nocturnal breeze.

Mountain dwellers attach great importance to these regularly alternating day and night breezes, which are to them a sign of fine weather. They have accordingly been given a host of local names, although they are, in fact, always due to the same cause. One of the most famous is the cold, dry *pontias* or *wind of Nyons*, descending from the icy summits and following the valley of the Eygues as far as Nyons. It rises shortly after sunset and drops at daybreak. In former times, the inhabitants of the valley, unable to find any explanation for the origin of the *pontias*, attributed it to a miracle. Later they thought that it was due to the cold air blowing from the large, deep fissures in the mountain of Devez. One day, according to a local legend, they decided to fill up the holes to put an end to this disagreeable wind, but the vegetation in the valley perished and they were finally obliged to re-open them!

In the Drôme Valley the rising current is called the *vésine*. In Basse-Savoie and Bas-Dauphiné, a mountain wind, particularly violent in the winter, is called *matinière*. The people living on the shores of the Lake of Geneva know the *bornan*, coming down from the valley of the Dranse and blowing as far as the centre of the Great Lake, as well as the *môlan* or *môlaine*, which comes from the Môle through the valley of the Arve and is nearly spent when it reaches Geneva. There is finally the cold *Joran* or *Jura wind*, which, during the summer months from May to October, rises suddenly in gusts, especially at the end of the day or after sunset. Coming down from the mountains at Nyon, it does not always reach the Lake but blows with great force over the villages at the foot of the Juras. It flattens the crops and is dangerous to navigation. It is also felt on the lakes of Neuchâtel and Bienne.

To the mountain breezes may be assimilated certain very strong currents of air which blow over the mountain passes in either direction and are due to the difference of atmospheric conditions and temperature between the two slopes. The *Maloja* wind in the High Engadine is an example. These mountain winds are well known to Alpine climbers and can, when they blow with great force, make some ascents difficult and others, which are arduous enough in calm weather, quite impossible. It is unlikely that an experienced Alpinist would be carried away by the wind, however violent it was, but it chills his body, endangers his balance and may seriously obstruct his vision in the swirling snow.

Cold Winds

Every district has its cold wind, which is particularly dreaded in countries of medium and high latitudes for the sudden drop in temperature and the snowstorms which it brings in winter and for the damage which it causes to crops. Such in France are the Alpine wind, the *mistral* in Provence and the *cantalaise*. This latter blows with extreme violence over the Plateau of Aubrac, accompanied by snowstorms which block the roads in winter. There is also the *soulèdre*, a north-east wind inimical to vegetation and felt keenly on the Grands Causses.

Italy suffers from the ravages of the *tramontana*. Rumania has the *crivetz* which, after sweeping over the immense snow-covered plains of Russia, in a few hours lowers the temperature by 20° C. in Bucharest. Belgrade owes its disagreeable winters to the *kochava*, a dry and very cold south-east wind, which makes certain summer residential quarters absolutely uninhabitable during the winter.

The North Wind

This very dry, cold wind is well known in western Switzerland; it is, in fact, so frequent and violent at Geneva, that it may be said to be one of the town's peculiarities. It is the prevailing wind over a large part of the Alps and is felt with varying degrees of severity as far as Mont Ventoux. It is also strong in northern Italy and is called there the *tramontana*.

This *north wind*, which comes from a direction varying between north and north-east and may locally even be north-west, is due to the presence of high-pressure areas to the north-west of the point where it is felt or to the presence of low-pressure areas to the south-east, or to a combination of these two causes. It is most frequent and attains its greatest force in winter and at the beginning of spring. At Geneva, on windy days, its speed is of the order of 45 m.p.h., sometimes even 60 m.p.h. Rushing down from the passes in the mountains, it raises blizzards and causes a drop in temperature of from 8° C. to 10° C. in twenty-four hours. It is this wind which makes the winter so severe in the Basin of Gap. "In the southern and not very high districts of the Alps," writes E. Bénévent, "the north wind makes the winter." In the summer, it is usually less violent and, although it sometimes produces a disagreeable sensation of cold, it does at least, at that season of the year, bring some delightfully cool days. As a general rule, the north wind brings fine weather and disperses the clouds. In the autumn, however, on the shores of the Lake of Geneva, it may be accompanied by a cloudy sky and beating rain. It is the *black north wind* which has given the inhabitants of Geneva the proverb, "With the *wind*[1] it often rains. When it rains with the *north wind* one is wetted to the skin."

In the French Alps, the north wind has been given a host of local names, all connected with the direction from which it blows. In the Haute-Maurienne, at Bonneval-sur-Arc, it is the *Iseran,* rushing down from the pass of that name and causing the houses with their heavy stone roofs to tremble, while at Lans-le-Bourg it is called the *Vanoise* (coming down from the Vanoise Massif). In the valley of the Guisane, as far as Briançon, it is called the *wind of the Galibier* and in the valley of the Clairée the *wind of the Rochilles,* while at Saint-Paul-sur-Ubayes it is the *wind of the Vars Pass.*

The north wind is responsible for a great deal of damage. Like all strong winds, it is capable of damaging roofs, removing tiles and chimney-tops and uprooting trees. On days when it is keen the Lake of Geneva is rough and stormy. In the winter it piles up snow in the railway cuttings, as often happens in the Lus Pass, and the trains are stopped completely for several hours. In other places it blows so violently that the roads become impassable, especially in the narrow valleys. Between Lus and Serres, where the valley of the Buëch narrows, the north wind sweeps along with redoubled force. On some days at the Plan-du-Rose, midway between Lus and Serres, movement on the roads is practically impossible. In many valleys of the southern Alps, especially towards Sisteron, in the Bas-Gapançais and in the Diois, the greatest complaint that is made against the north wind is not the disagreeable sensation of cold which accompanies it, even in the summer, but the fact that it threatens the fruit crop. At Puy-St.-Martin, in western Diois, the trees are all strongly bent towards the south and fruit trees will not grow at all. In the district of Gap their cultivation is limited to specially favoured places. Trees are not planted in the exposed fields, but there are orchards wherever the ground is a little sheltered, even in the smallest dales. In the past, there used to be vineyards on the slope of Cologny near to Geneva, but they have since been abandoned because,

[1] The *south-west wind* at Geneva is simply called "*the wind*".

being exposed to the north wind, their yield was very poor. Every year it causes more or less severe damage in the countryside round Geneva. It is particularly disastrous at the end of March and during April, for the frail shoots of the trees are mercilessly torn away and the blossom on the fruit trees is soon strewn over the ground in the orchards.

The north wind is not only harmful to crops on account of its force, but also because it dries them out and brings late frosts in spring. It is all the more dangerous to young plants when it occurs after a period of rain because, by promoting evaporation, it can have an intensely chilling effect. Even very severe cold is always less harmful to vegetation when the weather is calm because its drying action is then much less.

The Mistral

The *mistral* blows in Provence from the north or the north-west and has a considerable and generally very harmful influence. It rages principally in the spring and autumn, when the pressure is high in the north and west of France and a régime of low pressure exists over the Mediterranean between Spain and Italy. It is a local wind, cold, dry and very strong, and the relief of the land plays an important part in its origin. The air, moving southwards over a large area through Burgundy and the Saône Valley, finds, after it leaves Lyon, only a narrow corridor, becoming narrower still past Valence, between the Dauphiné Alps on the one hand and the Massif Central and the Cevennes on the other. It therefore follows exactly the course of the Rhône Valley, which, by canalizing it, greatly increases its force.

The mistral is felt all over Provence and may extend as far as Nice, but it is strongest in the Rhône Valley, especially between Valence and Arles. It seldom goes beyond the coasts of Provence and Languedoc. It may drop at the end of twenty-four hours, but, according to popular belief, it is usually felt for three, six and nine days consecutively. Precise records, however, do not confirm these assertions.

The violence of the mistral, which in some narrow gorges reaches a speed of over ninety miles an hour, had already been observed in ancient times. "The Crau," wrote Strabo, "is ravaged by the wind called '*melamboreas*'—an impetuous and terrible wind which displaces rocks, hurls men from their chariots, crushes their limbs and strips them of their clothes and arms." It scarcely seems to have abated since the time of Strabo, if we can believe the accounts of the damage which has been attributed to it. One story is quoted of a violent outburst of the mistral in a snowstorm on 3rd January, 1786, when some travellers and shepherds with their flocks all perished in the Crau. It seems that in the past, when this wind was blowing, the stage-coaches had to be secured with ropes to prevent them from overturning and being cast into the Rhône. This had been known to happen.

The mistral is capable of throwing a man from his horse, upsetting a cart-load of hay and even overturning the wagons in the railway station at Miramas. On one occasion a line of trucks which had been badly braked was pushed twenty-five miles without an engine by the mistral between Arles and St. Louis! It so often raises pebbles that for a time the windows on the north side of the Château of Grignan, not far from Montélimar, were left unglazed because they were being continually broken. It is also recorded that the Abbé Portalis was thrown by the mistral from the summit of the Mountain of St. Victoire and was killed by the fall. The lively imagination of the people of the south may have tended to exaggerate the force and the disastrous effects of the mistral; nevertheless the stories are not all unfounded. Its unleashed violence raises the dust on the roads, snaps off leaves and branches, shakes the fruit trees, sheds the grain before it has ripened and scatters the haystacks.

The rows of cypresses planted to break its force and protect the farms and growing crops are not always a sufficient obstacle to prevent it from lifting the tiles off the roofs of houses and making their doors and windows rattle. The mistral is active in spreading forest fires and sometimes causes such furious turbulence on the Rhône that boats are unable to go up the river. Sailors often have difficulty in steering their barges to prevent them from cutting across the current and striking against the pillar of a bridge. It is a serious handicap to shipping in the ports of Marseilles and Sète and may temporarily delay the arrival of ocean-going vessels.

The mistral is a cold wind which brings sudden changes of temperature; in fact, in twenty-four hours, it can lower the thermometer by 10° C. At Marseilles the coldest days in the winter are those when it is blowing, but it is most dangerous in the spring.

Sometimes it lasts for a week and it is, of course, inevitably hostile to vegetation. By drying up the soil—a serious matter in Provence, where the rains are seasonal and very irregular— and by carrying away the loam, it increases the aridity of the district. Trees in places particularly exposed to the mistral grow stunted and bent and are frequently broken, just as herbaceous plants are withered and torn up. To protect their crops and vegetable gardens, the farmers surround them with reed fences.

As compensation for the harm that it does, the benefits conferred by the mistral seem very limited. It assists the extraction of the salt in the salt-marshes of the Midi by promoting evaporation; it also gives some pleasantly cool days in the summer and there are people who praise its bracing quality. It would, of course, be an exaggeration to go so far as to claim that, although the effects of the mistral are disagreeable and even dangerous, its absence is often worse. That, however, is the gist of an old popular saying, which affirms that the town of Avignon is unpleasant when the wind is blowing and unhealthy when it is not:

> *Avenio ventosa*
> *Cum vento fastidiosa*
> *Sine vento venenosa*
> *Omni tempore odiosa.*

The Bora

The word *bora* is derived from the Slovenian, but it is also used by the Italians to designate the prevailing wind in the Adriatic, which, coming overland from a direction between north and east, blows over the Adriatic and Italy in the winter when the pressure over the Balkans is high. It is a descending

c

current, very cold and dry, and has its origin on the snow-covered Karst Plateau, where the temperature is very low, and on the Dinaric Alps—districts directly overlooking the sea.

The bora is felt particularly on the Istrian and Dalmatian coasts, from Trieste to Ragusa, where it falls in a cascade, bringing waves of terribly cold air. Its dryness makes it disastrous to vegetation and it has the effect of increasing the aridity of the district already caused by the permeable nature of the soil and the seasonal character of the rains.

The topographical conditions and the canalization which it undergoes through the mountains gives the bora in some places an extraordinary violence. Through the passes it is literally unleashed and on some leading into the interior it may stop the traffic for several days in succession. On one occasion it overturned a train at Klis not far from Split. In that district, the trains cannot run when the wind is very high.

Where a chain of mountains rises steeply along the coast, the bora reaches the sea with added force. On the Adriatic coast, islands such as Pago, a little to the north of Zara, disappear from view behind the screen of fine spray raised by the furious wind assailing their eastern shore and stirring up the foaming sea. In January, 1939, a large vessel in the course of construction at the dockyard at Trieste was burnt down in six hours while the bora was blowing furiously in a temperature of —15° C.

Although at certain points this wind is particularly formidable, other places have a reputation for being sheltered from it. Generally speaking, it rapidly loses its force in passing over the sea.

CHAPTER III

The Föhn

THE mountain dwellers in the Swiss Alps and the Tyrol are very familiar with this hot, down-driving, extremely dry and violent wind. It has an indisputable influence—sometimes harmful, sometimes beneficial—on the Alpine economy. Among the different types of local winds, it is certainly one of the most interesting to study.

The term *föhn*, generally used to designate it in Switzerland, seems to come, by contraction, from the Latin *favonius*, the wind of the Midi sung by Horace; hence are derived in Romanche, a language spoken in a part of Grisons, the words *favugn*, *favuogn*, *favoign* and *fogn*, the last being in use in the Livinenthal. Some authors, on the other hand, trace the origin of *föhn* to the old Gothic word *fôn* meaning "fire".

In Switzerland it also has several local names. At Glarus it is called the *funa* or *fün*; at St. Gall it is the *fö* and at Appenzell the *pfö*. On the Lake of Neuchâtel it usually blows from the south-east and is called the *uberre*, and at Fribourg it is the *white wind*. On the shores of the Lake of Geneva it is called the *Vaudaire*, a term which must come from the word *vallesaria*, used in the seventeenth century and signifying "wind of Valais". At Bouveret, a district situated on the eastern edge of the lake, it is still called the *wind of Valais*.

Origin and Characteristics of the Föhn

The *föhn* is felt particularly in the northern valleys of the Alps, where it usually comes from the south, sometimes from the south-east and more rarely from the south-west. The inhabitants of these valleys know its characteristics well and even if it blows for only a short time they never confuse it with any other wind. The police in the little Swiss cantons of Uri, Schwyz and Glarus, where it shows such fury, certainly have no need to consult a meteorologist before enforcing the regulation that all fires must be extinguished when the *föhn* is blowing. For a long time, however, its exact origin was completely misunderstood. The mountain dwellers thought that it was capable of holding in check any storms from the west and of keeping away the rain as long as it persisted. In fact, a popular saying promises fine weather while the *föhn* lasts and predicts that when it does stop it will fall in the mud:

> *Föhn macht schön*
> *Wann er vergoht*
> *Fallt er in Chot.*

("The *föhn* brings fine weather; when it disappears, it falls in the mud.")

In the Alpine valleys it often occurs at the same time as the rain-bearing wind from the south-west in the rest of Switzerland and the latter usually finishes by carrying the *föhn* away.

In the past some people thought of this wind as being peculiar to their valley, while others saw in it ". . . a current of hot air, which, rising over the burning sands of the Sahara and known to travellers in the desert by the name of the *simoom*, passes over the Mediterranean to exert its enervating and oppressive influence on the people of Italy, who curse it as the *sirocco*, and, crossing the Alps, arrives in Switzerland where the mountain dwellers bless its coming." This popular belief concerning the origin of the *föhn* is derived from an ingenious theory put forward by the geographer, Arnold Escher de la Linth, and reported by the Swiss writer, E. Rambert (Ref. 63), who devotes some interesting pages to the long discussions of scholars in the nineteenth century before they could agree on a rational explanation for it.

His study of the traces of ancient glaciers in Switzerland led Escher de la Linth to think that the creation of the Sahara, the beginning of the *föhn* and the amelioration of the Alpine climate must have occurred simultaneously and were directly connected one with the other. He considered that, before becoming a desert, the Sahara had been a sea swept by wet winds. Once it had emerged and dried up, the whole area became a furnace. The rising air, heated by contact with the burning sands, was always ready, if the circumstances were propitious, to blow in hot gusts towards the shores of Europe. The lower layers of air, charged with moisture after passing over the Mediterranean, produce in Italy the sirocco, which is stopped by the barrier of the Alps. The upper layers, on the other hand, not encountering any obstacle, may, at the slightest provocation rush northwards and into the valleys and thus give rise to the *föhn*. According to Escher de la Linth, the beginning of the *föhn* marked a change in the climate of the Alps. It was the end of the Ice Age and of Arctic temperatures in Switzerland. After that, the winter snows melted every summer up to altitudes of 6,000 and 9,000 feet and the glaciers disappeared except on very high ground.

This picturesque and ingenious hypothesis of Escher de la Linth, based on the assumption that the *föhn* was a fiery child of the Sahara that had freed the Alps from their snow, appealed for a time to popular imagination, but Dove, a Berlin meteorologist, did not take long to prove that the *föhn* could not possibly have an African origin.

E. de Martonne has well defined it as a current of air determined by the position of the great centres of atmospheric pressure but modified by the influence of the relief of the country. It coincides with low pressures to the north-west of Switzerland, generally with a violent storm over the Atlantic near the coasts of Europe, and with high pressures to the south of the Alps. During the great gale of 4th and 5th January, 1919, the difference of pressure between Basle and Lugano was 10.5 mm.; both these towns are at the same altitude, but one is situated 125 miles north of the other.

The *föhn* is produced by masses of air which ascend the southern slopes of the Alps. As it rises, the air, being submitted to steadily diminishing pressure, expands and consequently cools; it then jettisons its moisture in the form of rain or snow

according to the altitude and the season. The fall in temperature is about 1° C. for every 300 feet of altitude and when it
reaches the summit of the mountain the air, being dry, rushes
down into the valleys of the northern slopes. In its descent it is
heated up by compression, but, because of its dryness, the rise
in temperature is considerably greater than the fall caused by
loss of heat during its climb and it reaches the plain much
hotter than it was when it started on the other side. Between
the northern and southern slopes the difference may be 10° C.
to 12° C., while the moisture content is reduced by forty to
fifty per cent.

When the *föhn* is blowing strongly, heavy rains occur on
the high crests of the Alps, causing floods in the valley of the
Rhône, above the Lake of Geneva, in all the tributaries from
the south, while those from the north maintain their normal
level. The *föhn* would not occur if the air did not drop its
moisture as it climbed the slope because, if the water vapour
remained in suspension in the form of mist or clouds, there
would be no difference in temperature, at the same altitude,
on the two slopes of the mountain range.

Effects

The two essential features of the *föhn* are its extreme heat
and dryness; in addition it purifies the atmosphere and
produces exceptional visibility.

When this hot wind blows in the depths of winter, it is not
unusual for the temperature to rise suddenly by 10° C. or even
more. Früh (Ref. 32) mentions some Alpine districts where
temperatures of 12° C. to 17° C. above the normal have been
recorded. At Altdorf, during the storm of 1st to 3rd January, 1900,
the thermometer rose to 17.9° C. On an autumn day the
temperature may rise to that of the hottest summer days and
make some favoured Alpine valleys the mildest districts in
Europe. At the time of the *föhn* of 23rd to 25th September,
1866, it was hotter in Switzerland than at Lisbon and the
midday temperature at Zug on 24th September was higher
than any recorded in Italy on that day.

The temperature of the *föhn* has the peculiarity of being
absolutely independent of the sun. It is sometimes as oppressive

at midnight as it is in the middle of the day, and the shady side of the valleys where the wind is blowing is often as scorchingly hot as the side exposed to the sun.

Other effects are that the hair of some people becomes electrified; furniture, walls and woodwork in the houses crack and there is danger of fires arising from the slightest spark. The risk is all the greater because the houses in the Alpine villages are mostly wooden and fire can spread in a short time to a whole village or town. The *föhn* has been responsible for a number of fires in Switzerland, for example, at Glarus on 10th and 11th May, 1861, and at Meiringen, which has been destroyed twice, once on 10th February, 1879, and the second time on 25th October, 1891. For this reason, in all districts particularly exposed to it, very strict police regulations have been made to avoid the risk of fire when a storm occurs. Smoking is expressly forbidden and all fires have to be extinguished. People are not even allowed to cook their food. Special guards (*Föhnwächter*) are appointed by the communes to enforce these regulations which everyone regards as necessary and is anxious to obey.

The *föhn* has an effect on the nervous system so that many people feel ill at ease. There is a feverish quality about its intense heat which makes it difficult to bear. Plants and animals are also affected.

When it blows strongly, it can cause terrible damage, and, as soon as they feel its approach, the mountain dwellers hasten to gather in their flocks and herds. They extinguish all fires and the Alpine village closes in on itself as long as the storm lasts. Crossing the tops of the mountains, this down-driving wind beats down into the valleys on the northern flank of the Alps and then sweeps over the whole of the Swiss plateau, tearing the roofs off the chalets, flattening the crops, uprooting the trees and devastating the forests. The well-remembered gale of 5th January, 1919, literally laid waste the mountain forests of St. Gall and Appenzell. But, while the wind leaves devastation in some areas, others not far away miraculously escape. All the roofs of the houses in one village may be torn off without even the smoke rising from the chimneys in a nearby hamlet being deflected. Unharmed fields may be found quite close to ruined tracts of grass land.

Boatmen on some of the lakes where it rises without warning have good reason to fear the *föhn*, which is a great hindrance to navigation and is often the cause of accidents. Brunnen, on the Lake of Lucerne, has a special port for use on days when it is blowing. Sometimes it is so strong that even steamers have to stop to take shelter there. At Ouchy, on the Lake of Geneva, it has been necessary to build a special jetty, the "Vaudaire pier", to protect the port from this wind.

When the *föhn* blows in the winter, and more particularly in the spring, the avalanches in the mountains begin to move and the streams, suddenly swollen by the rapidly melting snows, bear away trees and enormous pieces of rock. These floods may cause severe damage.

Benefits of the Föhn

This wind has acquired the name of *Schneefresser* in the Alps because it is said to devour the snow. It is generally estimated that twenty-four hours of the *föhn* in the spring are equivalent to two weeks of sunshine. In some cases, undoubtedly, its astonishing power in this respect has been rather exaggerated. It is said, for instance, that it can melt twenty-three inches of snow in one day, and at Grindelwald it once did melt more than twenty inches in two hours, and in 1863 nearly twelve inches of ice covering a pond near Appenzell disappeared in a day. In Valteline, when the *föhn* rises on a spring evening, the people are in the habit of saying that "the wolf is going to eat the snow during the night".

Thus, although the *föhn* is a formidable wind capable of bringing disaster in its train, it is not without its advantages. Some people, forgetting its misdeeds, go as far as to say that it is the good spirit of Switzerland. They claim that without it the Alpine meadows would be covered with snow late into the spring and that the glaciers would grow and threaten to invade the valleys. Thanks to its beneficent action, however, the snow often disappears in the spring with great rapidity; the air turns soft and warm as if by magic and everything in nature is reborn and becomes verdant; the grass shoots forth and the cattle can at last leave their sheds. A saying in Grisons affirms that "if the *föhn* did not interfere, neither God nor his sunshine would ever be able to make an end of the winter snows".

Certainly it enables many narrow and rather sunless valleys, which could not otherwise be cultivated, to enjoy a milder climate. If it is late in coming, the crops are endangered. It is also said that in the autumn the *föhn* dries the second crop of hay, even in the shade, and that it ripens the fruit and the harvest. Without it, neither the maize nor the grapes in the cantons of St. Gall and Grisons in the upper valley of the Rhine would reach maturity. The southern type of vegetation that is found, in places, in Switzerland and in the Tyrol in the neighbourhood of Innsbruck is also attributed to it. Thus, Interlaken and Brienz, with their gardens of sub-tropical plants, enjoy the mild climate at the foot of the Bernese Alps and owe their privileged position to the fact that they are warmed by the *föhn*. It has its greatest effect in places which are sheltered from the north winds; this is the case at Gersau, Vitznau, and Wegnis on the northern shore of the Lake of Lucerne, which is called the "Swiss Riviera" and where one can see cypresses, laurels, camellias, fig trees and a number of exotic plants growing in the open air.

In spite of a tendency to exaggerate its beneficial effects, therefore, it certainly helps to free the mountains from their winter snow and gives a milder climate to certain valleys where it may blow more than a hundred times in a year.

Other Winds of the Föhn Type

The phenomenon of the *föhn* is found in many mountainous countries besides Switzerland and the Tyrolese valleys. It is, however, rare in the French Alps owing to the general orientation of the chain, whose folds are parallel to the direction of the south wind which gives rise to it. As Bénévent shows clearly, it could only be very localized there. It occurs, for example, in the eastern part of Bas-Dauphiné and Basse-Savoie if the wind of the Midi is forced to climb the foot-hills of the Alps. In the winter abnormally hot winds sometimes blow in the French Alps, but their high temperature is generally explained by their southern origin. They are accompanied by torrential rains, which cause avalanches and melt the snow.

The *Lombarde*, well known in Maurienne, is related to the

föhn. It is produced by masses of air which climb the Italian slope of the Alps and rush over to the French side in violent gusts. The *Lombarde* is usually a rather warm wind, but it can, at certain times, be very cold. It occurs when the atmosphere is very dry over the plain of the Po. No condensation then takes place while the air is rising and its temperature drops, by expansion, about 3° C. for every thousand feet of altitude. In the winter when the current rises slowly, the air is cooled still further by contact with the snow-covered slopes and reaches the passes at a very low temperature. The heating up which should be produced on its descent is partly nullified by the influence of the ground, so that, at the same altitude, the *Lombarde* is sometimes colder on the French side. On the other hand, when the air over the plain of the Po is hot and moist, heavy rains fall on the Italian side and the *Lombarde* in Maurienne is then a real *föhn*.

The *autan,* which blows over the whole of Haut-Languedoc and part of Aquitaine, is hot, dry and violent. It is caused by the approach of Atlantic depressions which reach France through the Gulf of Gascony, or by the existence of an anti-cyclone in the south of Central Europe. The autan always coincides with rainy weather on the coasts of Languedoc where the *marin* blows—a moist east wind which is generally not very strong. As it crosses the Montagne Noire and the Corbières, the characteristics of the *marin* are modified in a similar way to those of the *föhn* and it becomes the autan, which acquires tremendous force in passing the threshold of Lauragais between Carcassonne and Castelnaudary. A part of the road from Carcassonne to Mazamet, which is near Les Martyrs and known as the Roc Villain, is very dangerous in this wind and cars are sometimes overturned by it. The autan, on account of its dryness and turbulence, has a damaging effect on crops.

Winds, analogous by their origin and characteristics to the *föhn* in the Alps, are felt in the High Tatra Mountains (Carpathians), in Japan and in Iran. The north wind that is called *zonda* in Argentina belongs to this type. Scaetta (Ref. 69) mentioned the existence, to the west of the great dorsal chain of the Congo in Central Africa, of real *föhn* corridors charac-

terized by a local rise in temperature and a reduction in
rainfall.

In New Zealand the great *west winds* strike against the
Southern Alps, the large mountain range in the South Island.
The rainfall is heavy on the western slope and the glaciers go
down towards the sea amidst the majestic and perpetually
green southern forest. These winds have lost their moisture
when they reach the eastern slope of the Southern Alps, which
has a semi-arid aspect in strange contrast to the other side.
As the steppes and heath-land take the place of the great forest,
the wind on this side often has all the characteristics of a
true *föhn*. Such is the famous *north-wester* of the Canterbury
Plains, devouring the winter snow and swelling the river-beds
in a few hours.

Winds similar to the *föhn* are well known on the coasts of
Greenland, where sudden changes of temperature are recorded
in the winter; for example, at Scoresby Sound, the thermometer
has been known to rise in a short time from —20° C. to +4° C.
—a difference of 24° C. caused by these warm winds. This
phenomenon is particularly frequent along the west coast, and
at Jacobshaven a rise of 25° C. has occurred in twenty-four
hours. One Christmas Day +9° C. was recorded there, although the normal temperature at that time of the year is
only —12° C.

The Chinook

Chinook is the old name given by the Indians to a hot, dry,
west wind which blows in the valleys of the tributaries of the
Missouri. It is now used to designate the equally hot and dry
wind descending from the Rockies in the west or the south-
west and raising the temperature of the Middle West of the
United States and of the Canadian prairie on the edge of the
range. This is a true *föhn* and has an undeniable effect on the
climate of an extensive region to the east of the Rockies. The
winter isotherms are significantly higher towards the north
where the chinook is the prevailing wind. Thus, Calgary, at an
altitude of 3,390 feet is warmer by 9° C. in December and
10.8° C. in January than Winnipeg at only 820 feet. The
chinook makes the winter climate of Alberta one of the most

variable in the world. This province is alternately submitted to the influence of masses of icy air from the polar regions and to the warming effect of the Chinook winds. The latter are more frequent in the southern districts, where they quickly melt the snow on the prairies and enable the cattle to graze outdoors during most of the winter. The chinook may still be felt as far north as the district of the Peace River and it extends eastwards to the centre of Saskatchewan.

When it blows in the winter, it may produce in a few hours a rise in temperature of the order of 30° C. The thermometer may thus suddenly move from −20° C. to +10° C. In some winters the chinook prevails for a large part of the season and the average temperatures are then much higher than the normal. For example, C. C. Boughner (Ref. 9) cites the month of December, 1930, when the average temperature at Calgary was 0° C., although in December, 1933, it was only −15° C. In February, 1935, Calgary had an average temperature of −1° C., against −24° C. in February, 1936.

The chinook enables the limit of the cultivatable area to be pushed far into the north of Alberta and Saskatchewan. The northern limit for wheat, which hardly goes beyond the 49th parallel in eastern Canada, extends in Alberta to Lake Athabaska at the foot of the Rockies, at a latitude of sixty degrees north.

CHAPTER IV

Hurricane Winds

EVERY country has its own particular gales, which, while not in general attaining the force of the tropical cyclones, nevertheless cause great damage from time to time. The gales in the temperate zones, like those which occur especially during the winter and at the time of the equinoxes on the coasts of western Europe, take the form of a vortex having a low barometric pressure at the centre. They advance from west to east across the Atlantic, swept along by the great convection current of the west which prevails in medium latitudes. But, as L. Houllevigue (Ref. 36) writes, "The contour of the *gale* is not as regular as that of the *cyclone*. The isobars, or lines of equal pressure, instead of being concentric circles, are more irregular in shape. The depression itself forms a deep and wider cavity, often extending more than 600 miles in each direction. The circulation of the wind (which is always anti-clockwise in the northern hemisphere) is consequently slower, and that is why gales are less disastrous and less localized in their effects than cyclones."

France lies to the south of the path usually followed by the centre of these disturbances and to this she owes the rain-bearing *south-west* and *west* winds. When the wind turns to the north-west, it is a sign that the depression is moving past. The countries of northern Europe, situated to the north of the path of the depressions, are, on the other hand, exposed to the east and north-east winds, which are hot and dry in the summer and very cold in the winter.

The havoc caused by heavy gales in Europe is well known.

Every winter the newspapers have occasion to report it: ships lost, damage along the coast, trees torn up, buildings blown down, telegraph lines put out of order, and floods inland. Gales can have particularly serious consequences along low-lying coasts as the wind causes a heavy swell on the sea and makes it more dangerous. This is what happens on the Dutch coast when the *north-west* winds coincide with the equinoctial tides. Gales can also cause temporary variations of several feet in the level of the Baltic. When a hurricane from the east or the north-east sweeps over it, a mass of water is carried in the direction of the wind and accumulates in the pocket formed by the German coasts, the Danish Archipelago and the tip of Sweden. As the Sound and the Great Belt and Little Belt do not provide sufficient outlet for this water, it swells into an enormous wave and submerges part of the generally low lands forming an obstacle in its path. Ch. Rabot (Ref. 61) quotes the example of the terrible hurricane from the east of 12th to 14th September, 1872, which caused disastrous floods along the coasts of Mecklenburg, Schleswig Holstein and Denmark. The wave raised by the wind forced back the rivers and eroded, or left alluvial deposits on, vast tracts of land normally at a safe distance from the action of the waves. The same author also mentions the hurricane of 31st December, 1904, in the course of which the Baltic seaboard between the Pomeranian Bay and the coast of Jutland was submerged, as well as the south and east coasts of the Danish islands. Towns were flooded, ports damaged, agricultural land devastated and railway lines cut. In some places, only the dikes stopped the flow of the waters into the interior.

The *Pampero* in Argentina is a south west wind which comes from the Cordillera of the Andes, crosses the Pampas and is felt as far as the Rio de la Plata and the Brazilian coasts. This very strong, cold wind follows the route of the depressions moving from west to east.

The squalls of the *norte*—a menace to navigation—descend from the valley of the Mississippi and blow violently in the Gulf of Mexico. The coasts of Chile are exposed to these gales which are formidable in their violence, although some warning of their approach is given by the appearance of the sky and the fall of

the barometer. They have caused the loss of more than one
large sailing ship when these vessels came to load up with
nitrates and anchored on very exposed open roadsteads where
the wind raises a rough sea in a few moments.

The *burster* on the southern coast of Australia is a gale
caused by the passage of a depression moving from the south.
It gives rise to violent storms in the Great Australian Bight and
causes a sudden drop in the temperature at Melbourne; this is
all the more noticeable when the *southerly burster* succeeds the
hot burning wind from the interior.

The most dreaded winds in Siberia are the *bourans,* which
blow from the Arctic Ocean, bringing snowdrifts and intense
cold in the winter and sweeping furiously over the enormous
area of the Steppes.

The storms in western and equatorial Africa, which are
incorrectly called "tornadoes", bear no resemblance to the true
American tornadoes. They occur at the changes of the season
and are of short duration, though very violent. They consist
of heavy gales generally followed by torrential rains. The
approach of one of these storms is an unforgettable experience.
The atmosphere is oppressively heavy, without a breath of air,
the sky is inky black, streaked with flashes of lightning. Suddenly a light breeze rises, then the wind begins to blow furiously.
If one is surprised in the forest by a "tornado", the only thing
to do, unless there is a clearing close at hand, is to stand flat
against the nearest and stoutest-looking tree to avoid the falling
branches. It is, however, only a precarious shelter; the gigantic
trees of the equatorial forest, in spite of their apparent strength,
are generally not very firmly rooted, and many of them collapse
in a thunderous crash dragging their neighbours down with
them. The natives living in the great forests are well aware of
this risk, and, while it is the custom in other countries to use
trees as a protection against the wind, they prudently cut down
all those standing near the edges of their villages. The same
precaution is taken along the railway tracks. These African
"tornadoes" are a danger to aircraft. Again, the wind raises,
in a few moments, short waves with deep troughs in the rivers
and lagoons, so that, for example, as soon as they see these

waves approaching, the boatmen on the Congo rush to shelter near the bank.

Cyclones and Typhoons

The *typhoons* of the China Seas, and the *cyclones* of the Indian Ocean and the Antilles, are different names for tropical hurricanes. They are vast whirlwinds having an average diameter of 300 miles, following a parabolic path, increasing in size but losing their intensity as they proceed. The rotation of the wind in these whirlwinds has a constant direction in each hemisphere. North of the equator the movement of the air is anti-clockwise, but in the southern hemisphere it is the reverse.

The centre of a cyclone is seven to fifteen miles wide and is the most dangerous area for ships which have not been able to move out of its path. Caught in this centre, even the stoutest vessel is never certain of being able to extricate itself. The strength of the wind increases from the periphery to the vicinity of the centre, where an absolute calm reigns but where the sea is turbulent with enormous waves coming from all directions. On the two sides of the central area of calm the wind blows from diametrically opposite directions. The passage from the centre is short and the change round of the wind very violent. Fortunately, long and patient observations of cyclones have made it possible for navigators to determine the position of the centre and the path. They can recognize in what part of the cyclonic area they are situated, whether they are in the worst sector, called the "dangerous edge" —because the rotational speed is added to the speed of progression—where they run the risk of being dragged towards the centre by a wind speed of over sixty miles an hour, or, on the other hand, whether they are in the "manageable edge", where the speed is lower and the risk consequently less.

Cyclones were much more dangerous in the days of sailing ships than they have been since the coming of steamers and the use of radio, which enables coastal meteorological stations to warn ships at sea. The latter are obliged by international regulations to give a wireless signal to the nearest coastal port if they meet a cyclone or even if they merely observe the premonitory signs.

Most cyclones, in fact, are preceded by characteristic indications at least twenty-four hours in advance. The atmosphere is abnormally clear and calm and the visibility unusually good. There is a heavy swell on the sea and the sky at sunset is tinged with copper. The most certain indication, however, is given by the barometer. Usually in the tropics its movement is very regular but its daily rhythm is interrupted before it starts to fall at the approach of a cyclone. The weather becomes steadily worse until the gale begins. Dense, black hurricane clouds appear on the horizon. When the centre is approaching, the rain begins to fall in torrents, because a cyclonic depression is a huge funnel of rising air which expands and therefore cools in its ascent and causes its water vapour to condense. As long as the hurricane lasts, the foaming crest of the waves and the clouds skimming over the sea are indescribably mingled in a yellow fog of salt water.

The average speed of a cyclone's forward movement is fifteen miles an hour, although some, especially the typhoons of the China Seas, have an extremely slow motion. The bad weather which accompanies them may last two days or a little longer, but the time during which the wind is really unleashed does not generally exceed twelve to fifteen hours. As a rule, the maximum intensity is of short duration and the most violent gusts, which precede or follow the passage of the calm zone in the centre and are responsible for most of the damage caused by cyclones, sometimes last only a few moments.

In both hemispheres, their season begins after the summer solstice. In the Atlantic it is from July to October, the worst month being September. In the northern part of the Indian Ocean they are particularly frequent from September to November, when the north-east monsoon succeeds the south-west.

There are no tropical storms in the eastern parts of the oceans; they are, in fact, unknown along the west coasts of Africa, America* and Australia. Those in the Antilles arise at the centre of a triangle of which the Azores, the Cape Verde Islands and the Antilles constitute the three vertices. The

---

*Editor's Note.—Nevertheless, the west coast of California used to be notorious for its sudden storms.

D

cyclones passing over Madagascar begin to the east of the Rodrigues Islands and those which sweep the Melanesian archipelagos and the coasts of Queensland in the South Pacific have their origin in the region of Fiji. The typhoons of the China Seas start round the Philippines.

Ravages Caused by Cyclones

Tropical hurricanes have been responsible for thousands of victims as well as enormous material destruction. The damage caused by the worst storms in Europe gives only a small idea of the disastrous effects of the typhoons unleashed on the China Seas or of the cyclones which sweep over the Antilles. Small boats are tossed like straws on the waves and are either totally engulfed or driven ashore. Sailing ships are dismasted. Even large vessels are in grave danger. Boats in the ports loose their moorings and are torn away from the quays. Often they collide and sink or are cast ashore by a tidal wave.

It is estimated that a wind of ninety miles an hour exerts a pressure of 21 lb. per square foot.* This pressure is 84 lb. per square foot with a wind speed of 180 miles an hour, which may well be attained by the strongest gusts in a cyclone. Pressures of several thousand pounds are therefore exerted on surfaces of a few square feet, which are often quite unable to withstand such a force. Houses collapse on their occupants; sheets of corrugated iron fly in the air like paper. Whole towns are sometimes razed to the ground; this happened some years ago at Tamatave and Ponte-à-Pitre, which in a few minutes took on the pitiable aspect of a battle-field. Electric lines are blown down, trees uprooted and crops totally destroyed.

The effects of the wind and the salt spray soon deprive a tropical island, covered with luxuriant vegetation, of its verdant appearance. When a cyclone was passing over the New Hebrides I remember seeing thickly wooded islands assume in a few hours a curious wintry aspect. All the leaves had been torn off their branches by the wind or reddened by the salt, while a large number of trees lay on the ground.

Translator's Note.—The pressures of winds of different velocities on buildings and various types of structure are given in a comprehensive British Standard Code of Practice, CP3 *Code of Functional Requirements of Buildings*, Chapter V (1952), "Loading", published by the Council for Codes of Practice for Buildings.

During a cyclone, the waters of rivers are often forced back up-stream. The torrential rains then cause dangerous floods in which bridges are carried away and roads submerged. Some coastal towns may also be inundated by tidal waves.

To quote only a few examples of cyclones which were real calamities, the hurricane at Cuba in 1844 sank about seventy ships and the damage caused in the space of a few hours was, at Havana alone, estimated then at more than £800,000. The cyclone of 10th October, 1870, was particularly destructive. It swept over the Antilles and then moved towards the North Atlantic. It sank about a hundred ships, tore up coral reefs from the bed of the sea, overturned solid buildings and on some islands left nothing standing, neither trees nor habitations. At St. Lucia, 6,000 persons were buried beneath the débris, while at St. Pierre on Martinique there were over 9,000 victims.

Coral islands, being low and narrow, suffer particularly in cyclones. A raging sea may easily submerge them, carrying away both vegetation and inhabitants. That happened in the Tuamotu Archipelago during a cyclone in 1903, when waves forty feet high swept over the islands of Marokau, Hao and Hikueru, washing away the coconut palms on which the inhabitants had taken refuge. Fortunately cyclones are rare in Polynesia, otherwise groups of coral reefs like the Tuamotu Archipelago would be practically uninhabitable.

In September, 1930, the Republic of Santo Domingo was attacked by a hurricane of unprecedented violence, which devastated the capital, Trujillo. The number of dead was estimated at 4,000 and the material damage at 30,000,000 dollars. The first two hours of the cyclone were particularly terrifying. It destroyed everything in its path—buildings, plantations and forests. The wind was then blowing at a speed of 150 miles per hour.

Every year tropical cyclones cause violent storms, especially in the month of September, along the coasts of the United States, as well as in the Gulf of Mexico and on the Atlantic seaboard. Fortunately their effects are not felt very far inland. The storm of 8th September, 1900, at Galveston (Texas) caused 6,000 deaths and damage estimated at 30,000,000 dollars. Two particularly destructive storms on the coasts of Florida occurred on 17th-18th September, 1926, and 16th-17th September, 1928. The first caused the death of 243 persons and the damage amounted to 76,000,000 dollars; in the second there were 2,000 victims and 25,000,000 dollars worth of damage. On 21st September, 1938, a tropical hurricane of indescribable violence assailed all the coastal states from New Jersey to Massachusetts. Four hundred and seventy-four persons lost their lives and the damage was estimated at 300,000,000 dollars.

Have cyclones any advantage to offer in compensation for so much destruction? One might be tempted to think so in reading these lines of Peltier, quoted by H. Marié-Davy (Ref. 50): "These convulsions of Nature," he assures us, "appear necessary to restore the equilibrium in the atmosphere by effectively mixing all its strata, so that, in spite of the terrors that they inspire, the inhabitants of the countries where they rage so furiously often pray for their coming." It must be admitted, H. Marié-Davy adds, that even if this advantage is of any real importance it is bought very dearly. Commander Bridet, the author of a study, published in the last century, on hurricanes in the southern hemisphere, considers the question from quite a different point of view. A chapter of his book is entitled: "A method of making use of cyclones to reach one's destination". He tells us that, as soon as sailors know the laws of cyclones, they have nothing to fear from them. In the days of sailing ships some of the more hardy seamen would plunge straight into one to speed them on certain voyages. English sailors call it "taking a ride on a cyclone". On 24th October, 1842, Captain Miller of the *Lady Clifford* was able, by taking advantage of a cyclone whose centre was passing over Pondicherry, to sail in a remarkably short time from Nagore to Madras. In the same way, Captain Erskine, in July, 1848, considerably reduced the time taken on the voyage from the Cape of Good Hope to Sydney.

Forecasting Cyclones

The meteorological stations that have been set up in tropical areas to give warning of the approach of cyclones and to study their course have been of great service both to sailors and to the local inhabitants. Warned in time, navigators can modify their course to avoid the path of the cyclone and the civil population can take precautions or evacuate their homes to seek refuge in special shelters. Thanks to these measures, the cyclone which struck the Bahamas in September, 1929, caused only a small number of casualties, compared with the hundreds of deaths which often occurred as a result of hurricanes in the highly populated islands of the Antilles. The Kanakas in the New Hebrides, especially those living on the island of Tanna, where the ancient high coral reefs have many spacious grottos, take shelter in these natural caves when a a cyclone occurs.

Special stations for forecasting cyclones have been established at Antananarivo in Madagascar, on the Willis Group of islands off the coast of Queensland, at Phou-lien near Haiphong and at the Observatory of Zi-Ka-Wei near the French Concession of Shanghai. In the year 1930 alone, the last-named station sent out warnings of fifty-one depressions, twelve gales, and thirty-four typhoons—a total of ninety-seven —and dispatched more than 18,000 telegrams. In the same year, it received 64,365 telegraphed messages bringing information from various meteorological posts and ships of all nationalities sailing in the China Seas. This observatory at Shanghai, where typhoons occur in the summer months from July to October and are most frequent in August, has forecast the approach and followed the course of more than 1,000 over a period of sixty years. Without it, many ships would have been lost.

Tornadoes

The *tornado* is the most violent and most spectacular meteorological phenomenon. The wind speed on its immediate path is greater than that in the worst tropical cyclone. The true domain of tornadoes is the eastern part of the Great Central Plain of America; the States most affected are Texas, Oklahoma, Kansas, Arkansas and Iowa. There tornadoes are a typically American phenomenon. In no other region in the world do they find conditions so favourable to their formation and nowhere else are they so frequent, violent and destructive.

They accompany the great thunderstorms of the spring and summer, which are also typically American phenomena.

A tornado is a strong eddy current of air moving, nearly always east or north-east, at a speed of thirty to forty miles an hour, but the speed of rotation at the interior many be as much as 400 to 500 miles an hour. This vertiginous whirlwind approaches with a peculiar whistling sound which soon becomes a deafening and terrifying roar. Fortunately the area affected is narrow, usually not more than a quarter of a mile. It has been noticed with surprise that an extremely small distance, of sometimes only a few yards, may separate a district which has been devastated by the passage of a tornado from another which has remained unharmed.

J. B. Kincer quotes some statistics to show the enormous destruction caused by tornadoes in the United States. In the twenty years between 1916 and 1935 there were 2,800 in which 5,224 persons lost their lives and the material damage amounted to 230,000,000 dollars. Three hundred and seventy-five of these tornadoes were responsible for 100,000 dollars' worth of damage each and thirty for more than a million. The following list gives details of some of the worst disasters:

Place	Date	Deaths
Erie (Pennsylvania)	26th July, 1875	134
Louisville (Kentucky)	27th March, 1890	106
St. Louis (Missouri)[1]	27th May, 1896	306
La Sandusky (Ohio)[1]	1924	
States of Missouri, Illinois, and Indiana[1]	18th March, 1925	689
Gainesville (Georgia)[1]	6th April, 1936	203

There are between 100 and 200 tornadoes annually in the United States and the average number of victims is 150 to 400.

Tornadoes are not, however, entirely unknown in Europe,* but there they are more often called whirlwinds or sometimes

[1]Damage estimated at more than 10,000,000 dollars.

*Translator's Note.—An interesting article describing the tornado at Tibshelf in Derbyshire on 19th May, 1952, appeared in the issue of Weather for July, 1952. The authors, F. A. Barnes and C. A. M. King, mention several other tornadoes which have occurred in England.

cyclones. They have a very strong suction effect and can raise heavy objects and carry them some distance away. At Moncetz, in 1874, even men and animals were lifted several yards from the ground. At Hallsberg, in 1875, a whirlwind threw a threshing machine over the ruins of a barn. The worst tornado in France was at St. Claude in Jura. This occurred on 19th August, 1890, and travelled forty miles between St. Claude and Romainmotier in the Swiss Juras in thirty-seven minutes, passing through Les Rousses and the Valley of Joux. Everything in its path, which varied in width from about half a mile to two miles, was devastated.

Although some whirlwinds have an extremely small diameter of about 30 to 160 feet, they can still be very destructive, owing to the rapid rotation of the column of air. The violent tornado which swept over the district of Ruchfeld near Basle on 24th July, 1939, affected an area only about 100 to 160 feet wide. It was reported that the wind turned sometimes to the right, sometimes to the left; it blew the roofs off houses and uprooted trees of two feet in diameter. It raised over three feet of water from a pond and threw it on to neighbouring fields.

Blizzards and Snowdrifts

Blizzards, especially in mountainous districts and in high latitudes, have on more than one occasion been fatal to travellers who have lost their way in the blinding snow and finally perished of cold.

The worst blizzards are those in the Antarctic and they are the principal characteristic of this continent, which is the windiest in the world. The approach of a blizzard is usually very sudden; in a few moments a calm may give way to a wind of thirty to forty miles an hour, which soon increases in speed. The gusts driving clouds of fine snow before them make everything invisible. When the hurricane is at its worst, it is impossible to distinguish a man or a tent at a distance of a few yards. Any object falling to the ground is immediately carried away. Observations have consistently shown that in a blizzard the wind keeps the same direction and usually drops without changing it. The gusts are strongest when the storm is nearly over. Sir Douglas Mawson, the Australian explorer, christened

Adelie Land, where he stayed from 1912 to 1914, "The Home of the Blizzard".* On 2nd July, 1913, he recorded a wind speed of eighty-five miles an hour at Cape Denison. After dropping to fifty-one, it increased to 116. Mawson remarked that the wind maintained an average speed of 107 miles an hour for eight hours. Gusts approaching 200 miles an hour were indicated by the anemometers used. The explorer's camp could not have withstood these hurricanes had it not been buried in the snow.

It was thought, until Scott's last voyage, that the winds from the Pole—and this is the direction from which the blizzards come—were comparable to the *föhn* in Greenland because the temperature in the winter during a blizzard is generally abnormally high—higher than during a calm period. Dr. Simpson, however, showed that this increase in temperature was not due to a *föhn* effect, but to the fact that in calm weather there is, in the Antarctic, an inversion of temperature. Between the ground and an altitude of 1,600 to 3,200 feet the temperature may rise about a dozen degrees Centigrade. The lightest wind increases it by mixing the cold lower layer of air and the upper layer which is appreciably warmer.

Snowdrifts are well known to the inhabitants of Newfoundland and Canada, who are accustomed to long and rigorous winters. While an icy wind is blowing, the fine, dry, whirling snow fills the air and penetrates everywhere. The famous "spray" storms of the islands of St. Pierre and Miquelon are brought by the very strong, dry north-east winds. Several are recorded every winter, but, according to the islanders, they have been less frequent and violent in the last twenty years. They are, however, still capable, at times, of forming enormous piles of snow in the streets of St. Pierre, where trenches have to be dug to make a path to and from the houses.

The French Canadians have adopted the expression "powder" for these storms and say that it "powders" when the north-west wind raises eddies of fine snow which in some places settle in enormous heaps. In the west of Canada they are called *prairie blizzards*.

Very strong gales called *bouranes* on the coasts of Siberia

* *Translator's Note.—The Home of the Blizzard*, by Sir Douglas Mawson. Hodder and Stoughton, London, E.C.4.

bring large quantities of snow. The *pourga*, during the winter, blows over the icy tundra in the Kamchatka Peninsula, sweeping away tents and sleighs. The swirling snow, which may be as fine as flour, or large-grained and crystalline, pierces the skin and is sometimes so dense that any object a yard away is indistinguishable. In such a wind it is impossible to open one's eyes and difficult to breathe and to keep on one's feet. A temperature of 10° C. to 15° C. below zero usually accompanies the *pourga*. When it is damp and the temperature is about 0° C., those who have experienced it assure us that this kind of hurricane is still more formidable. The *damp pourga*, fortunately rare, penetrates the thickest clothes in a few moments and makes the cold intolerable.

CHAPTER V

Wind as a Determining Factor of Climate

GENERALLY speaking, and particularly in the medium latitudes which correspond to the temperate zones in the two hemispheres, every change in the wind is followed by a more or less marked change in the weather. Ordinarily, the farther the wind has travelled, the more marked is this change. The great atmospheric currents to a certain extent bring with them the climate of the countries they have just passed over.

The wind, alternately bringing cold and heat, rain and drought, is a primary factor in the climate and it may be said that everywhere the successive predominance of winds determines the salient features of the atmospheric régime.

Thermal Role of the Wind

The influence of the wind on the variations of the thermometer is well known. When, after a period of damp and mild west wind, it turns to the east, it brings great heat in the summer and severe cold in the winter. These contrasts are still more marked in North America, where continental influences are more pronounced. In the United States and Canada, very hot south-west winds, which make the atmosphere stifling in the large towns, sometimes suddenly alternate with icy breezes from the north, bringing late frosts in the spring and early frosts at the end of the summer. The Australians justifiably dislike the *hot winds*, which turn the cities of Adelaide, Melbourne and Sydney into veritable furnaces on certain days in the summer and cause a general exodus to the higher and fresher resorts of the Blue Mountains and the Australian Alps as well as to Tasmania and New Zealand. The "heat waves" of Australia coming from the overheated desert regions of the interior and making the thermometer rise to 40° C. envelope the towns in a shroud of yellow dust and cause a large number of deaths. They are usually followed almost immediately by the *southerly burster*—a south wind which brings a sudden drop in temperature and glacial rain.

All countries in the temperate zones are subject to these sudden changes of temperature caused by a change in the direction of the wind.

In many places, the winds, far more than the seasons, control the variations of the thermometer. The thermal role of the wind is especially noticeable on islands having a cold climate. In Newfoundland, the temperature is essentially dependent on the force and the direction of the winds. So it is on the small subantarctic islands, where the slightest change in this direction immediately has an effect on the thermometer.

The influence of the wind is particularly remarkable at San Francisco, a city which enjoys the freshest climate in the United States. The average temperature for July is only 14° C. and during the whole period from May to August the thermometer varies between 11° and 17° C. These abnormally low and even temperatures are due to the action on the sea of the west winds which drive before them the upper layers of the water and cause the deeper waters, which are exceptionally fresh for this latitude, to rise to the surface.

The principal reason for the extremely cold winter at Verkhoyansk in Siberia, where the thermometer falls to 50° C. below zero and even lower and which was for a long time considered the coldest place in the world,[1] is the almost absolute stillness of the air. This district, sheltered from the winds by the rather high surrounding terrain, occupies a basin in which the cold air accumulates. The slightest movement of

[1]This record now belongs to the station of Oymekon.
Translator's Note.—The claim for Oymekon (or Oymyakon) is based on comparative observations at both places for the four years between 1930 and 1933. (See *The Cold Pole of the World*, by Terence Armstrong, *Weather*, Vol. VII, No. 12, December, 1952.) Oymekon lies in a frost hollow on the upper Indigirka, about 400 miles south-east of Verkhoyansk; it stands at an altitude of 2,625 feet, as compared with 400 feet for Verkhoyansk.

the air is immediately followed by a rise in temperature. Well to the north of Verkhoyansk, the tundra extending along the shore of the Arctic Ocean is exposed to violent winds, yet the winter temperatures there are appreciably higher. In these extreme northern regions of Siberia the coldest winds in the winter blow from the south-east, while the north winds are known as warm winds.

The Humidity Cycle

The atmosphere contains a variable proportion of water vapour and the essential function of the wind is to carry it from one place to another, to lift it from the sea and precipitate it on the land. The humidity cycle therefore depends on the wind. In this respect, the air currents prevailing at the earth's surface can be divided into two groups. The first comprises the *evaporation winds,* which become saturated with moisture in passing over the sea; the second corresponds to the *rain-bearing* winds, which deposit this moisture in the form of rain or snow. A type belonging to the first group is the trade wind, which, blowing in the direction of the equator and becoming steadily warmer as it travels, evaporates a much larger quantity of water than it deposits. The general winds, *south-west* and *north-west* respectively in the temperate zones of the northern and southern hemispheres, belong to the second group and become steadily cooler as they approach the Poles.

Besides this general circulation of moisture caused by the great atmospheric currents, there are more local causes which affect the rain régime in a number of places. High mountains near an enclosed sea encourage rainfall. For example, a large part of the moisture evaporated from the Caspian Sea during the hot summers is carried by the north winds to fall as rain on the northern side of the Elburz Mountains, where it waters the forests, pastures and fertile lands of the provinces of Guilan and Mazanderan.

The wind régime regulates the distribution of rainfall everywhere and makes the difference between maritime countries, where the prevailing winds generally coming from the sea are saturated with moisture, and the interior of continents, where the winds, after a long journey overland, are dry. In western

Europe, the humidity is controlled by the prevailing *south-west winds*, which are saturated after crossing the Atlantic.

"When it reaches our shores, this air, being forced to ascend, cools by expansion; there is therefore less rain on the sea than on the coast and less on the coast than on the hills, plateaux or mountains which rise farther inland. But as it moves eastward, the wind loses its moisture; therefore, behind a littoral zone of increasing humidity, the land becomes steadily drier" (L. Houllevigue).

The wind régime is also responsible for the existence of the two desert zones which encircle the earth. It is, in fact, the persistence of the trade winds, engendered by the sub-tropical high pressures, which has resulted in the sterility of the Sahara, of Arabia, of North-West India, of Arizona and, south of the Equator, of the Kalahari Desert, of Northern Chile and of a large part of Australia. Blowing over continents, the trade winds are dry.

Over the sea they also behave like dry winds unless they encounter an obstacle such as a mountainous island or coast, like the Serra de Santa-Marta in Brazil, which, by causing part of their moisture to condense, produces heavy but localized rains on the slope which the winds strike directly.

Most mountainous tropical islands which are subject to the régime of the trade winds have two distinct climates, one on each side of the range of mountains. The side struck by the wind, the north-east or the south-east according to whether the island is situated to the north or south of the equator, receives torrential rains. The other side is dry. The sharp contrast in the vegetation on the opposite slopes, due to the difference in rainfall, clearly shows the effects of these two climates. The same wind may promote luxuriant growth to the east, while a few miles to the west it allows nothing to grow but a few stunted bushes. Thus, under a sky often dark with heavy clouds, the coasts and slopes of the islands in the New Hebrides which are struck by the south-east trade wind are covered with thick and verdant forests. To the west and the north-west, where the sky is much clearer, stretch vast undulating prairies, with here and there a few acacias bent by the wind, and a sort of scrub-land in the most southern islands. These differences are found in New Caledonia, in the Fiji Islands, in the Hawaiian Islands and in nearly all the archipelagos of the Pacific.

Madagascar also presents very dissimilar climates. The east coast, receiving the trade wind, is characterized by an evergreen type of vegetation, while on the west coast the plants are deciduous and are even xerophytic in the desert areas of the south-west. The agricultural possibilities for each of these climates are, of course, very different.

These local differences, caused by the orientation of the mountain slopes in relation to the trade wind, are characteristic of a large number of tropical islands and have brought into use the expression "windward side" to denote the wet and relatively cool part of the coast which directly receives the wind and where the sea is generally rough, and "leeward side" to denote the drier and warmer sheltered part where the sea is safer for navigation.

Réunion has accordingly been divided into two administrative areas, the east being the more fertile windward district, and the west the leeward district where rain is rare. A distinction of the same kind is found in some neighbouring archipelagos. The Society Islands are divided into "Windward Islands" (Tahiti and Moorea) and "Leeward Islands "(Raiatea, Borabora, etc.). In the Lesser Antilles, there are the Windward Islands (from St. Thomas to Trinidad) which are directly exposed to the wind, and the Leeward Islands (from Blanquilla to Oruba) which receive it after it has passed over the first group.

It is as a factor of humidity or drought that the wind has the most direct effect on the human race. This close relation between man and the wind appears in a striking fashion in southern Asia, where the monsoon distributes both the rainfall and the population. In India, the seasonal occupations as well as the prosperity of the different regions vary according to their situation in respect of this wind. Bertoqui says that the wet summer monsoon approaches India from the south-west and the north-east at the same time. The south-west current strikes the Western Ghats where the rain falls in torrents (78 to 157 inches) and then, sweeping across the interior plateau, becomes so dry that at Madras it can be compared to the sirocco. The other current, corresponding to the Bengal monsoon, is stopped by the Himalayas and produces still more abundant rains, which make Bengal and Assam the wettest regions in the world (Cherra Punji: 590 to 787 inches of rainfall annually). The Bengal monsoon spends its moisture as it goes up the valley of the Ganges so that when it reaches the plain of the Indus it has become completely dry. The middle and lower basins of the Indus, which are not in the path of either branch of the monsoon, are therefore perpetually arid. They form the Thar or Indian Desert, the development of which is being attempted through irrigation.

Influence of Relief

In some places, the wind is felt with such unusual frequency and force that their climate gains an unfavourable reputation. Generally, this disadvantage is due to the local topography. The configuration of the land has a marked effect on the wind régime. In some cases it favours the progress of atmospheric currents, in others it diverts or completely obstructs them.

Many towns are famous for their extremely windy climate. Baku on the Caspian is nicknamed the "Hub of the Winds". The record undoubtedly is held by Wellington, the capital of New Zealand, of which it is said that its inhabitants can be recognized anywhere by their instinctive gesture of lifting their hands to hold on their hats when they turn the corner of a street!

Other places are in a privileged position and are relatively sheltered. A favourable disposition of the terrain, in arresting certain winds, can cause a considerable local improvement in the climate. What a contrast, for example, between Geneva, situated in a corridor where the wind sweeps along between the Salève and the Juras, and Montreux, at the other end of the Lake, which is well protected by high mountains and enjoys very mild winters! The high barrier of the Elburz Mountains, intercepting the winds from the north, makes the winters in Tehran easily bearable. Similarly, the southern coast of the Crimea, protected from the cold north winds by the prolongation of the chain of the Caucasus, is considered to have one of the most agreeable climates in the world. Alsace, sheltered from the west winds by the Vosges and from the east winds by the Black Forest, is open only to the winds from the north and south, which give it comparatively hard winters and hot summers, when tobacco can be cultivated.

A mountainous terrain is therefore, to some extent, a screen and many places situated at the foot of high mountains are relatively protected from the wind coming from that side. It would not, however, be correct to say that mountains are always an absolute obstacle to the movement of the air. If that were so, the Alpine valleys in Switzerland would not experience the *föhn* which climbs the crests. The Haute-Maurienne, often called the "Country of the Wind", would be sheltered from the Lombarde which crosses the mountains at the frontier, and the mistral would not blow on the Côte d'Azur. The French Alpine villages, leaning back against the northern slopes of the valleys and facing towards the south, ought to be well protected from the icy north wind, but many of them are only relatively shielded. The relief can sometimes serve as a protection against the wind; it breaks the power of the atmospheric currents locally and canalizes them, but it does not stop them entirely. Often, indeed, by obstructing the flow of the air, it only increases its force through openings in the barrier.

It is a well-known fact that the speed of a current of air is increased when the width of its passage is diminished. A wind with a moderate speed in a broad, open plain becomes much stronger when it passes through a valley or a gorge. A valley naturally facilitates the movement of the air in its own direction, so that the narrower the valley, the stronger the wind will be. Its effects are therefore generally greatest in valleys oriented in the direction of the prevailing wind. E. Bénévent mentions, in this connexion, the valley of the Grand Buëch, running north-south between Lus-La-Croix-Haute and La Faurie, where the very frequent north wind is sometimes so strong that a pedestrian walking against it is stopped short at every gust.

Salubrity of the Climate

There is a definite connexion between the salubrity of a district and the natural conditions which ensure its ventilation.

Places which are situated near to each other but not oriented in the same way in relation to the prevailing winds, or subject in one case and not in the other to the influence of a local breeze, often have very different climatic characteristics. As a general rule, large plains, provided that they are not swampy, and extensive plateaux are healthy. Masses of air with nothing to hinder their progress often move more rapidly over the latter than round many open hill-tops.

Narrow coastal plains, on the other hand, although they face the sea breezes, may be badly ventilated if they are dominated by a high range of mountains. The east coast of Madagascar, the Brazilian seaboard from Rio to Bahia, the Atlantic coasts of Central America, the Choco region at the foot of the western slopes of the Andes in Colombia and the Pacific coast of America all suffer from this disadvantage and their climate cannot be considered as very salubrious.

There is no doubt that certain tropical countries where the air is very still are unhealthy, but it would seem that the benefits of the wind and its influence on the salubrity of the climate were formerly rather exaggerated. Studying the role of winds in hot climates, R. Radau (Ref. 62) and Ch. Pauly (Ref. 59) considered them of prime importance as purifiers of the atmosphere. In their opinion, the wind régime, the quantity of pure air, rich in oxygen, brought from the sea by the atmospheric currents and whether the winds were stopped or allowed to pass by the configuration of the ground when they reached the coasts, played an essential part in making a region healthy or unhealthy. The great endemic diseases of malaria, cholera and yellow fever were due to the stagnation of the atmosphere; the centres of miasmata were nearly always found in enclosed valleys where the air was not renewed sufficiently often, while places of exceptional salubrity like the plains of La Plata and of Paraguay were constantly swept by the wind.

Ch. Pauly observed that Algeria abounds in badly ventilated, unhealthy districts, where the mortality was high shortly after the French conquest. The same conditions exist on the east coast of Spain, "where the intimate connexion between climatology and endemic diseases is revealed most clearly". In fact, the coastal mountains which dominate the Mediter-

E

ranean shore from Tarifa to the Pyrenees make a succession of low-lying areas with almost tropical temperatures. The district of Murcia is a striking example. When Ch. Pauly made these observations eighty years ago the whole of this coast of Spain was a hotbed of malaria. "When particular meteorological conditions," he said, "are combined with the local relief of the country, epidemics occur like the terrible outbreak of yellow fever which decimated Barcelona in 1821." The Catalonian capital is encircled by mountains on all sides except the east, which faces the sea. During the epidemic of 1831 a very light breeze blew constantly from the south.

One cannot, however, read without a smile some of the accounts of the dangers attributed to the stagnation of the atmosphere. They may be judged by these few lines written in the last century by Lind about a bay in the island of Roatan off the coast of Honduras. "The ships anchor in a basin surrounded by high mountains and inaccessible to the winds. The air becomes so stagnant that, after breathing it for a few days, one is suddenly attacked by violent vomiting, headaches and delirium and blood starts to ooze from the pores of the skin. The sea water would probably putrefy in such places if it were not kept in motion by the ocean currents."

The wind certainly has a beneficial effect, but it should not, of course, be exaggerated. Ch. Pauly and many others praise the bracing quality of the trade wind and the agreeable climate that it brings to the South Sea Islands, but some over which it blows very keenly are far from being entirely salubrious. In the New Hebrides, where it is particularly violent, malaria is extremely widespread, even along the coasts directly exposed to the wind, and in the south of the group the very low-lying little island of Aniwa is literally infested with mosquitoes. They also abound in New Caledonia, in the Fiji and Samoa Islands, at Tahiti and in a large number of archipelagos in the Pacific, where they are in some places a real plague. The trade wind seems to have no effect on the legions of anopheles, culices or stegomyas.

The proportion of days when there is a sea breeze greatly affects the popularity of health and bathing resorts, the most popular being those which are the least windy.

CHAPTER VI

Physiological Effects of the Wind

THE wind has an indisputable influence on the human organism and its effect appears in many different forms. Sometimes it acts directly, in a mechanical way, exerting a pressure which is the obvious cause of the fatigue that we feel when we are exposed to it for some time. At other times, its action is indirect and is a result of the variations in temperature and relative humidity which it produces. Finally, as a medium of transport, the wind carries and disseminates germs or solid particles, either suspended in the air or raised from the ground, which, when propagated, give rise to epidemics.

It was observed, in very early times, that atmospheric phenomena, particularly the wind, had a certain influence on human beings, from the psychological as well as the physiological point of view. Hippocrates, in his treatise, *Of Airs, Waters and Places,* had already drawn attention to the effect of the wind on certain ailments. "In towns frequently exposed to winds, such as those which blow between the east and the west, and which are sheltered from the north winds," he wrote, "the slightest cause can change sores into ulcers. The inhabitants lack force and vigour, the women are sickly and voluntarily barren, the children are attacked with convulsions or sacral disease, the men are subject to dysentery, to long fevers in the winter. . . ." The successors of Hippocrates, however, ignored the relationship between meteorology and medicine, which he had perceived, and did not take into account atmospheric influences on the health of their patients, although popular belief has always held that the wind, one of the essential features of climate and one of the easiest to observe, has a very marked action on the human organism.

It was only at the beginning of the twentieth century that a scientific study was made of the wind's physiological effect, and numerous researches, particularly, in France, those of A. Aimes (Ref. 2), L. Savignon (Ref. 68), G. Mouriquand (Ref. 55), and A. Missenard (Ref. 53), showed that not all popular beliefs concerning its influence were to be rejected. Precise observations have proved that this influence was evident in many cases.

The foundations of this new science, which sets out to study the relationship between medicine and meteorology and to which the name of "meteoropathology" has been given, have been very well explained in a recent article by M. Pasteur Vallery-Radot (Ref. 58), who very justly observes how delicate the question is, for the effects of different meteorological factors on the human organism are not produced in isolation. Usually they act simultaneously and it is often difficult to distinguish the cause of a particular phenomenon.

Influence of the Wind of the Midi

Dr. Mouriquand has made a special study of the pathological effects produced by the *wind of the Midi* in the Lyons district, a wind which results in a considerable rise of temperature and a reduction of the relative humidity. It causes in some people discomfort, headaches, rheumatic pains, a stifling sensation and a certain irritability, as if there were electricity in the air. It may bring on fits in epileptics, blood spitting in consumptives and attacks of asthma in those afflicted with the complaint. Some sickly children are habitually feverish when this wind is blowing. Babies, especially those recovering from serious infections, are particularly sensitive to it, and it may occasionally produce such a dangerous condition, through loss of moisture, as to result in death.

Dr. Mouriquand observes that, especially among children, but also in some adults, ailments appear which obviously coincide with meteorological phenomena but which cannot always be attributed with certainty to them. This applies particularly to attacks of asthma in children when the wind of the Midi is blowing. He adds that not all human beings are sensitive to atmospheric disturbances unless the latter occur with exceptional force. Those who are not influenced by these factors are called *meteorostabiles*. There are, on the other hand, some individuals, although only a very small number, who are extremely sensitive to meteorological variations and who feel them even before they are recorded by precision instruments.

Some interesting cases of such physiological presentiments, obviously connected with this wind in the Lyons district but also produced by other winds, such as the *föhn* in the Alps and the *Levanter* in Tangiers, have been observed in certain persons, who were very sensitive to physical changes in the atmosphere, from twelve to twenty-four hours, or sometimes even two days, before the wind started to blow. The effects have generally taken the form of disturbances, especially nervous excitement, sometimes intense, attacks of asthma or of rheumatism. These frequently disappeared when the wind rose.

Action of the Levanter

Missenard reports the singular effects, studied at Tangiers by Remlinger and Burnier, of this east wind which blows through the Strait of Gibraltar from the Mediterranean to the Atlantic Ocean. It causes, in some people living on the shores of the Strait, headaches, a sensation of oppression and distress, an inaptitude for work and a certain irritability of temperament. It opens lesions and brings on bouts of fever in tubercular patients and it has an obvious disturbing action on the highly-strung. Its effects are particularly marked in the summer.

Unlike the *föhn* and the *wind of the Midi*, the *Levanter* brings a fall in temperature. The variations of pressure which accompany it are equivalent to those of an ordinary fall and the relative humidity diminishes only slightly. As we have just seen, the troubles occasioned by this wind very distinctly precede its occurrence, except, apparently, in those suffering from tuberculosis. Remlinger has tried to explain these physiological presentiments by attributing them to the variation of the electric field and the ionization of the atmosphere. In support of this theory, he mentions the fact that the wind very often occurs at a considerable altitude about twenty-four hours before it is felt on the ground. This explanation is not, however, universally accepted.

In the opinion of Missenard, the hypothesis that the effect of winds is due to modification of atmospheric electricity is based partly on the analogy between the ailments associated with it and those preceding thunderstorms, when electric fields vary considerably, but he says that this explanation is generally not sufficient and will not bear a thorough examination. At the moment, therefore, we can do no more than state the undeniable fact that some human beings feel in advance the approach of a certain wind, as, for instance, the Levanter.

Effects of the Sirocco

This hot wind is well known for the depressing effect it has on human beings as well as on animals and plants. E. F. Gautier (Ref. 33) stresses its peculiar odour, which effects the mucous membrane so that the slightest breath of the *sirocco* is immediately recognizable It is a very dry smell and—although such cases are extremely rare and are mentioned only by way of curiosity—can have a toxic action on the human organism.

Many evils are attributed to the sirocco in North Africa. It causes attacks of cerebral excitement in soldiers stationed in lonely outposts and has been known to occasion outbreaks of suicide in the Foreign Legion.

In Italy the sirocco is moist after crossing the Mediterranean; in other places it is scorchingly dry. When it is blowing, the temperature remains, night and day, at about 40° C. As the air is dry, adults can bear the great heat fairly easily, provided that they drink adequate quantities of liquid to counteract the rapid evaporation from the skin. Very young children cannot support this excessive evaporation so well, and there is a high infant mortality from what is called "heat stroke". In less than two days a baby may die from complete loss of moisture. Sometimes the sweat glands cannot function actively enough to counteract the effects of a hot dry wind blowing at a high speed, and the temperature of the skin and of the body gradually rise to the level of the ambient air, causing a fatal heat stroke.

Effects similar to those of the sirocco can be observed in all countries subject to very hot, dry winds.

Various Pathogenetic Winds

Many other winds also have a clearly discernible action on the human organism. In fact, it may be said that there is no part of the world where the inhabitants do not consider that some winds have a more or less injurious effect on their health.

Everybody knows that draughts, damp evening air and cold winds in general are dangerous, often all the more so when they are least perceptible, for then fewer precautions are taken against them. There is, however, still a risk of contracting a chill through the sudden changes of temperature which they provoke. The cold breeze which comes down from the Sierra de Guaderrama is reputed in Madrid to be the cause of many cases of pneumonia, although it is too gentle to blow out a candle. As a Spanish proverb says, *Un aire sutil que mata a un hombre y non apaga un candil.*

The wind is sometimes blamed for its irritating effect on human beings, like the icy wind at Geneva, or for causing a general feeling of uneasiness like the *föhn* in the Alps, as well as for nervous tension and headaches. Huntington quotes the case of an ordinarily calm child who climbed howling to the top of a tree when a certain wind was blowing. The famous *120-days wind*, which blows northwards over the eastern part of Persia, has an irritating effect on the inhabitants and deprives them of all initiative and energy.

In many districts, certain winds are thought to cause lassitude, insomnia and various complaints. A curious fact is that it is generally east winds that are blamed. In this connexion, there is an interesting passage from a letter written by Voltaire at the end of the eighteenth century, in which he describes the effects of the *east wind* on the population of London. This clearly shows the psychological influence of the wind in certain cases.

"This east wind," wrote Voltaire, "is responsible for numerous cases of suicide. A famous court physician, to whom I confessed my surprise, told me that I was wrong to be astonished, that I should see many other things in November and March, that then dozens of people hanged themselves, that nearly everybody was ill in those two seasons and that black melancholy spread over the whole nation, for it was then that the east wind blew most constantly. 'This wind is the ruin of our island. Even the animals suffer from it and have a dejected air. Men who are strong enough to preserve their health in this accursed wind at least lose their good humour. Everyone at that time wears a grim expression and is inclined to make desperate decisions. It was literally in an east wind that Charles I was beheaded and James II deposed.' "

An old English proverb says that people, like animals, can expect nothing good from the east wind:

> When the wind is in the east
> 'tis good for neither man nor beast.

Winds have their part in popular legends, but they cannot be blamed for all the ills attributed to them. It is hardly likely, for instance, as was at one time believed at Poligny, that the *montaine*, a cold wind coming down morning and evening through the Vaux Pass from the Juras, can be injurious to the teeth.

Wind and Cold

The chilling action of the wind is a well-known phenomenon. The higher the speed, the colder is the current of air, at the same temperature and humidity. A current of air at 20° C. saturated with moisture and moving at a speed of nearly seven miles an hour, is as cold as still air at 14° C. Missenard, from whom this example is taken, stresses the fact that the thermal effect of the wind is very different at different seasons of the year. In the winter, when the temperature is low, the wind lowers the resistance of the human body; on the other hand, it increases it in the summer when the ambient temperature is too high. The interesting observations made by Huntington in various districts of the United States, particularly in New York, Washington and Chicago, show that the wind increases mortality in the cold weather and reduces it during heat waves.

In Siberia, where winter temperatures sometimes fall to —50° C., it is a well-known fact that the excessive cold can be borne with relative ease only because there is no wind. A still cold of —35° C. is more bearable than —20° C. in a wind. It is therefore the calmness of the air which enables the inhabitants of Northern Siberia to endure the intense cold. In fact, a popular proverb says, "No wind, no cold". As soon as the slightest breeze rises, it "cuts like a knife". For the same reason, the long winters of Newfoundland, where the wind is very constant and violent, are particularly rigorous, although the thermometer seldom falls below —15° C.

In the Antarctic, the wind is a vital factor of the climate from the human point of view. Every explorer has remarked that even very severe cold, at temperatures of —20° C. or —30° C., does not seriously affect the human body but that it can be very dangerous when the wind is blowing. Dr. Simpson has also shown that the terrible snowstorms in the Antarctic, during which visibility is reduced to a few yards, have an influence on the nervous system. Although the temperature in the winter is considerably higher in a blizzard than in periods of calm, in spite of the fact that the wind comes from the Pole, it would be sheer madness to attempt to travel through the fine, powdery snow. It has, in fact, been observed that men, surprised by a blizzard, lose the faculty of clear thinking and become like wild creatures. This singular effect perhaps explains why Scott and his unhappy companions, after travelling more than 625 miles, died only 12½ miles away from a provisioning base. If they had been able to reach it, they would doubtless have been saved.

The Wind as an Agent in Spreading Infection

Apart from its climatological action to which the human body is sensitive, the wind disseminates pathogenetic germs deposited on the ground or suspended in the air and may be instrumental in spreading epidemics. It was observed in the past, during certain outbreaks of cholera, that air currents could carry the disease over short distances, as it appeared in villages situated down wind from an infected area.

The wind detaches germs from their resting place by a

purely mechanical action and it is capable of effecting real microbial separations. The outbreak of cholera on the shores of the Bosphorous in 1908, when the disease was transmitted by spray, is an example of germs being snatched up by the wind blowing over the sea. The wind generally increases the microbial content of the atmosphere by raising dust settled on the ground. Therefore, although the wind purifies the air in some districts by dispersing germs, it may also take part in spreading infectious diseases.

By scattering grains of pollen in the air, it helps to spread hay fever, which in many places during the summer is so much dreaded.

CHAPTER VII

Adapting Dwellings to the Wind

THERE are, in the construction of rural houses, as A. Demangeon (Ref. 27) very justly remarks, adaptations to the climate which evoke the idea of geographical determination. Undoubtedly, none of the different features of climate has influenced man in the building of his house more than the wind. Usually, in fact, it imposes the choice of a site and dictates the form and arrangement. It is generally to fight against the troublesome effects of the wind, rather than to take advantage of its benefits, that man has used his ingenuity to create, in very exposed places, a habitation specially adapted to stand up to squalls. The types of these dwellings vary considerably from one country to another and very often even in neighbouring regions.

The evolution of the house in the course of the ages, the use of more resistant and impervious materials and the improvement of methods of heating in cold countries have often produced a more solid construction than in the past. Man today has therefore partially escaped from the tyranny of the wind. In nearly every country there are old houses which enable us to judge what defensive measures and precautions had to be taken against it—precautions that have by no means been universally abandoned.

Situation and Orientation

The wind, in certain places, drives back and, so to speak, chases away, human habitations, and in this connexion Jean Brunhes (Ref. 11) cites the example of the almost deserted valley of the Reuss high in the Swiss Alps, where the *föhn*, often with terrible effects, blows so violently that most of the villages are huddled in the lateral valleys which are much more sheltered.

In many districts, the fear of the wind and the love of sunshine determine the choice of the site and the orientation of the rural home. There are few places where people do not complain of the severity of a particular wind, blowing more often and more violently than the others and bringing cold or heat, rain or drought. Thus, in very windy places, they seek out even the slightest shelter and the most favourable orientation for their homes. On very wind-swept coasts and plains, the houses naturally crouch behind whatever protection the country affords. Often they stand close together to offer more resistance to the storm, like, for example, the small and low granite houses of the town of Sein. Valleys are not necessarily wind corridors and in all mountainous districts there are some which, owing to their orientation, are relatively sheltered. This happens when they are perpendicular to the prevailing winds, and communities of people seek them for preference.

People everywhere try to avoid as much as possible the wind that they fear most. In Provence, most of the houses have a general north-south orientation with a slight inclination to the east so that the front is sheltered from the cold mistral coming from the north-west. In Languedoc, as the autan (S.-E.) is as troublesome as the north-east wind, the houses face due south, thus being exposed only at the sides. The vine-growers in the district of Orleans seek an orientation midway between the east and the south-east, thus reconciling their love of sunshine with their aversion to the south-west wind, that "cursed vile sou'wester" or that "vile wind", as they usually call it. In the Basque country, the houses are often built at the top or on the slopes of hills, generally facing east to be better shielded from the wet winds of the Atlantic, against which they are further protected by long roofs reaching almost down to the ground. The Flemish house, with its unsymmetrical roof covered with cemented tiles, turns its back to the north wind and, hidden behind the dunes, faces inland.

Almost every part of the world offers similar examples. Many houses in Annam (Central Vietnam) face towards the south to afford protection against the winds from the north, which are cold and unpleasant in the winter. Very often, too, the villages are shielded by small plantations of trees on the north side, not only for comfort but also for superstitious reasons.

In a large part of Madagascar the construction of the native houses is influenced by the altitude and direction of the prevailing trade wind, which drops warm, torrential rain all along the east coast and which, after being cooled by ascending the slopes leading to the high plateaux in the centre, brings damp, fog and cold. The houses follow the only conceivable plan by turning their backs to the bad weather and that is why the dwellings of the Hova tribe always face west. G. Julien (Ref. 39) reports that this rule was observed to such an extent that when the chiefs addressed a crowd of their followers they began, "All of us who face the setting sun". If anyone, for some reason or to satisfy a whim, had oriented his house towards the north, south or east, he would have been pitied and treated as mad, for to act in such a way would be to defy destiny.

In Timbuktu the houses are never turned to face east or west, the directions of the prevailing winds, for fear that evil spirits should enter them.

Nevertheless, man does not always take the precaution of placing his house on a favourable site to protect himself from the wind and its orientation sometimes leaves much to be desired. In some cases, of course, this disadvantage is offset by an advantage of another kind connected with the occupation of the inhabitants. In the past, some farmers on the north coast of France—and they may even do it today—were in the habit of storing their milk and butter in cellars opening to the north because they not only provided shelter from the sun but also admitted the sea breezes which were thought to bring some of the salt from the sea.

The situation of Miquelon, a fishing village off the coast of Newfoundland, may also seem surprising. Built on a narrow beach and exposed to all the winds, it has already been flooded several times by the sea. The choice of this rather hazardous site is explained by the ease with which the fishermen of Miquelon can put their dories out to sea and haul them in at

night. The absence of bays and the rocky nature of the coast
would make this difficult if the village were situated in a more
sheltered place.

Finally there are cases where man deliberately seeks an airy
situation for his home, this choice usually being imposed for
health reasons. Leenhardt (Ref. 46) remarked that the Kanaka
of New Caledonia does not want a sheltered place for his cabin.
"The sweet fresh air blowing in pleasant surroundings is for
him a symbol of well-being and is part of his psychological
climate." A customary leave-taking among these islanders is,
"Good-bye, we shall sit together again in the caress of the
wind."

Many Europeans living in hot countries try to build their
houses on high ground or in an exposed position to take advantage of any breeze which will lower the temperature and chase
away the mosquitoes.

Shape and Protection of the House

Men have always been obliged to protect themselves from
the severity of the wind and their first means of defence was
undoubtedly to adapt their habitation to it.

The most elementary type of dwelling, used by some tribes
in the southern hemisphere, is a simple screen placed as a
shield against the prevailing wind. The Australian aborigines,
not having any fixed abode, are satisfied with a temporary
shelter consisting of a screen of intertwined branches. The Ona
people of Patagonia erect a screen by stretching llama skins
over six rods arranged in a semi-circle and dug slightly obliquely
into the ground. The aborigines of Tasmania, now extinct, had
a similar shelter, but they used strips of bark instead of skins.

The hut, also an extremely primitive form of habitation,
was conceived mainly as a defence against the wind. Many
hastily constructed dwellings in use among peoples living in
very windy districts show how careful they are to shield themselves as much as possible from draughts. The nomads of Central
Asia make their huts completely draught-proof and fix them
firmly in the ground so that they cannot be carried away by the
wind. Eskimoes generally use skin tents in the summer, but
for the winter they construct dome-shaped huts with blocks of

ice cemented with water, which freezes as soon as it is poured
between them. The inhabitants have to crawl along a snow
tunnel and enter through a narrow opening made flush with
the ground. Inside, they are well protected from cold and
draughts as the only window is a transparent block of ice. The
Eskimo igloo, owing to its hemispherical shape, runs very little
risk of being overturned in a storm.

The nomadic tribes living in, or on the borders of, the
desert areas of French Somaliland, which is subject to very
violent winds, use low dome-shaped tents made with rush-
matting fixed on arches buried in the soil. Sometimes a stone
wall shields them on the most exposed side. While I was
travelling there, I remember meeting a tribe on the flanks of
Mount Gouda, who, instead of using a tent, were camping in
the shelter of a crescent-shaped wall about as tall as a man.
It was made of twisted branches and blocks of lava.

In Europe it is mainly in coastal and mountainous regions
and on great plains and plateaux that buildings show the
most obvious signs of being prepared for the onslaughts of the
wind. It is not unusual to see houses which are so low that they
seem to crouch on the ground. France offers an astonishing
variety of types, often showing by their shape and construction that they have been designed to resist the wind. "The fear
of hurricanes in districts where they occur frequently," writes
A. de Foville (Ref. 30), "is seen, not only in the orientation,
but even in the architecture of the buildings. First of all, there
is only a ground floor to give the wind less chance of taking hold
of the building." He adds that the low houses along the
Channel coast can give the impression of a country subject to
earthquakes.

At St. Pierre and Miquelon, which are exceptionally stormy
islands, the winds are sometimes warm and moisture-laden,
sometimes dry and icy, but they blow furiously whatever their
direction. The traditional house, which still exists in fairly
large numbers, especially at Miquelon, is small and low, consisting simply of a ground floor with no protrusions, and is
made entirely of wood covered with shingles.

In other places, not only is the building low, but the further
precaution is taken of having practically no opening on the

side turned towards the wind most to be feared. This is done in
Provence as a defence against the mistral. Often the windows
fit badly, especially when the mistral is blowing, because the
wood shrinks and makes them loose in their frames. Doors
and windows are therefore placed on the south side, which is
the only wall to be made of brick, the others being of stone.
The Provençal house is generally built with its length in the
same direction as the wind to lessen the risk of its being overturned. In Normandy, the most exposed wall is usually rein-
forced or faced with protective material. In many districts, it is
almost completely eliminated and replaced by a section of the
roof reaching nearly to the ground. This can also be seen in the
Dutch and Flemish farms, weighed down under their large
roof which shields them from the north wind, and with all
their windows opening on the side facing the sun. The Basque
etche is protected in the same way from the moist west winds,
and a very similar arrangement is found in the Moyenne-
Garonne, where the houses are also protected from the prevailing west or rain-bearing wind. The side facing west is called the
ploutzal and is shielded by a large roof in the form of a hood.
The windows open on the sheltered side called the *celat* (hidden).

The extremely low houses in the Vivarais (a region of
Languedoc) on the bare plateaux swept by the *burle* are
scrupulously oriented to the south, but on the north side not
even a part of their wall is exposed because the roof reaches
down to the ground. Some of the oldest of them are half-buried
for further protection. This arrangement recalls that adopted
by the Arctic peoples living on the Tundra. For the rigorous
winter, when they experience terrible snowstorms or *pourgas*,
they build themselves very low huts nearly sinking into the
ground. Some of them are made almost subterranean to run
less risk of being blown over in a storm, and consist of a ditch
with only the dome-shaped roof, made of planks and branches,
rising above the level of the ground. The entrance is through a
hole in the roof. In other cases in the Arctic, the thickness of
the walls is the only guarantee against the wind and the incle-
ment weather. The dwelling is then a low cone-shaped or hemispherical hut made of poles and branches covered with earth,
moss and bark. These huts have only a very narrow opening

A dust devil on the desert plain of Hanleh (Somaliland)—a column of sand raised by a gyrating wind similar to a waterspout raised from the sea

Sand ridges formed by the wind on the coast of Somaliland at Balambale

A sandstone rock eroded by wind-driven dust at Rombalds Moor,
Yorkshire

Wind channels cut through snow on the Isle of St. Pierre,
Newfoundland

A road cut through loess, Quito, Andes

A farm in the Juras with its south-west side protected by "tavaillons"

and look as though they have been flattened on the ground. The inhabitants of the Kamchatka Peninsula living on the shores of the Bering Sea also used to spend the winter in subterranean houses.

A very common precaution in many districts is to make a special device to protect the house against squalls and particularly against the rain or snow brought by the wind. For this purpose, various types of covering are used, generally placed on the side struck directly by the wind. Wood, in the form of shingles, slate, slabs of fibrous cement and sheets of zinc are the most usual materials.

Many houses in Aveyron, Tarn and Lozère are protected in this way by tiles of shale or slate placed on the west side, and sometimes on the south side, too. Some of the houses between Mazamet and Saint-Pons are surrounded by them. Both on the Swiss and French slopes of the Jura Mountains the same precaution is taken to preserve the walls of the houses from the destructive action of the rain and snow, which would increase the damage caused by the wind of the Midi (what is called the wind of the Midi in France or the wind of Geneva on the Swiss side is not really a south wind but the southwest wind which distributes rain all over western Europe). The protected side of the mountain dwelling is covered with shingles and small slats of fir wood partially overlapping, to which the local name of *tavaillons* is given. The front door is never placed on this side, which is called the *talvane*, and the windows there are shielded by a small hood, which is not found on the other sides of the house. Some old farms, still to be seen in the Joux Valley between La Cure and Le Brassus, particularly near Bois d'Amont, also have their roofs covered with *tavaillons*, which in the course of time take on a pleasant silver tint.

There are also constructional devices such as static aspirators, which are placed on chimney-tops to draw up the smoke, whatever the direction of the wind, and prevent it from being driven back when the movement of the air is downward. In countries accustomed to hard winters double-glazed windows are used as a defence against cold winds. In Newfoundland it is usual to place an extra door in front of the entrance to prevent

F

the snow from penetrating inside the house during a gale. This is a sort of portable wooden sentry box, which is generally removed in the summer. It has two side doors and the occupants of the house enter or leave by the one on the sheltered side. There used to be a similar custom at Cayeux-sur-Mer, in the Département of Somme, where the houses were provided with two entrances on opposite sides, so that if one happened to be blocked by sand from the neighbouring dunes the occupants could use the other.

It is much more rare, at least in Europe, to see part of a building specially adapted to take advantage of the wind.* The open barns discovered in the Alpine village of St. Véran, which, at an altitude of nearly 7,000 feet, is the highest on the Continent, are an example of this. They consist of a boarded floor under the roof, protected by a balcony. As the winter begins early and the farmers are obliged to harvest the crop before it reaches maturity, they place the sheaves on these boards to ripen in the wind.

There are some places where the wind is so strong that the houses have to be braced with guys. The buildings on South Georgia and the Kerguelen Islands, and on the small islands in the Southern Ocean frequented by whale fishers and seal hunters, could not stand up to the gales without this reinforcement.

In districts subject to cyclones, the enormous pressures exerted by the wind raise special problems, which Cattala brings to the notice of architects. He remarks that the colonial houses of the old style, with their large verandas, were particularly unsuitable. The solid cubic type of structure made of concrete, which was adopted for the reconstruction of Tamatave after the cyclone of 1927, is much better. These flat-roofed houses are perhaps not very pleasing to the eye, but they can stand up to hurricanes.

A building, even with its doors and windows closed, is never hermetically sealed, especially when there are openings for ventilation. The effect of a strong wind in a cyclone is to

Translator's Note.—The ruins of King Solomon's smelting works at the southern end of the Long Rift Valley, formed by the Jordan Valley and the Wadi Araba, show that these were built with openings north and south to take advantage of the constant winds there from the north.

make the pressure inside the building lower than the external static pressure. It is dangerous, therefore, to open a door or a window during a cyclone. The house should be closed as tightly as possible and solid shutters placed in front of all the openings.

The Roof

In many countries subject to high winds, special measures have to be taken to strengthen the roofs of houses. It is a fairly common custom to have steeply sloping roofs, but this is by no means a general rule, and in some very windy districts having a high rainfall or heavy snowstorms the roofs of the houses are only very slightly inclined. In eastern Canada and New-foundland, where the wind is violent and the snowfall heavy, the rural house is made of wood covered with shingles and generally has a roof in two sections, inclined at an angle of forty-five degrees and scarcely protruding beyond the wall. By the side of this old type of house, there are some modern houses with an almost flat roof.

In mountainous districts, and in many valleys in Piedmont and in the French Alps, where the wind would be strong enough to blow off the roofs, they are often covered with large stone slabs, set firmly in solid wooden frames which have been cut in the neighbouring forest.* These heavy roofs always have a very slight slope. In the Massif Central, the slabs are called *laves*, although they are not always of volcanic origin. The large schistose slabs which are used in many places in the Alps are called *lauzes* (also written *loses*). Stone roofs of this kind are so heavy that the houses often seem to be crushed under their weight and so low that they can almost be touched by hand.† Chalets and *mazots*, used as barns, may be seen in many valleys in the Swiss Alps. These buildings have very flattened roofs, made of planks of wood which are often piled with heavy stones as a precaution against the effects of the wind.

---

* *Translator's Note.*—The thatched roofs of old cottages on the west coasts of Ireland and Scotland are held down by ropes weighted at the ends with stones or tied to small projections built into the walls for the purpose.

† *Translator's Note.*—The cottages in the Orkney Islands frequently have only a ground floor and their roofs are made of thick stone slabs, from two to three feet square, which cannot be carried away by the wind.

At Elne in Pyrénées Orientales, on the side of Vermeille where the tramontana blows with such fury, special flat tiles are used, fixed at their upper and lower ends to prevent them from being blown off by squalls. In the Peninsula of Cotentin (Manche), the roofs are cemented to increase their solidity.

As Cattala points out, roofs were traditionally designed to support pressure when it is, in fact, suction which is often more dangerous. This is particularly the case in tropical countries where buildings are generally rather light. Corrugated iron roofing which is frequently used, for example in the South Sea Islands, is unsuitable in tropical regions exposed to cyclones. It is undoubtedly economical, but, apart from the discomfort that it causes in the house when it becomes overheated, its light weight and slight slope make it extremely dangerous in a hurricane when the sheets of iron may easily be blown off and carried some distance away. To avoid this danger, the colonists in the New Hebrides often place large stones or sandbags on the roofs of their houses or outbuildings.*

The flat roof, adopted since time immemorial all round the Mediterranean, is more suitable for hot countries than the sloping roof as it offers more resistance to very violent winds.

Ventilation

Houses, of course, have to be ventilated as well as protected against the dangers of the wind. Good ventilation is, in fact, the sovereign remedy for dampness and ensures the renewal of vitiated air.

In cold climates, great care is generally taken to exclude draughts but even in the most enclosed buildings there is inevitably a certain amount of ventilation due to the difference in temperature between the air inside and the air outside. The outside air always penetrates inside through some crack in the walls, even when the doors and windows are tightly shut, because a heated house in the winter is a centre of attraction for the cold air. This spontaneous ventilation depends on the degree of movement of the air outside, the number and size

---

Translator's Note.—Sheets of corrugated iron are frequently blown away in Orkney gales with such force that they are sometimes wrapped round the telegraph poles as they strike against them.

of the doors and windows and the condition and permeability of the walls and chimneys. It is greatly increased by the use of porous materials in construction.

These various factors generally help to maintain adequate natural ventilation inside the house, reducing the stagnation of the air and preventing it from reaching a dangerous degree of vitiation due to a high proportion of carbon dioxide. It has been observed that kitchens are always better ventilated if they are placed on the down side of the prevailing wind.

Natural ventilation, however, even if the windows are opened when necessary, may not always be sufficient, especially in large rooms containing a number of people. Opening the windows may only partially renew the air if the weather is calm. Artificial ventilation is therefore necessary in buildings intended to hold a large number of people. Besides complete aeration, it gives an almost constant temperature, by introducing fresh air in summer and heated air in winter. It is produced by suction with the help of draught chimneys or by fans.

Ventilation is particularly important in hot countries, where its principal purpose is to lower the temperature. For this reason, the shady Arab house with its narrow flagged courtyard, although it has no outside windows, is well designed to promote the free circulation of the air.

Methods of ventilation vary in different climates. In hot and very humid areas, especially near the sea or in the great tropical forest, the diurnal variations of temperature are very slight, often only of the order of 3° to 5° C. The best procedure there is to ventilate the building as much as possible by creating a continuous draught, either naturally by having several large openings or with the help of fans. The "punka", which has been used in India for a very long time and consists of a linen cloth hung from the ceiling, is sometimes preferred to the fan. It is worked by a native, who moves it to and fro by means of a cord, and it gives a gentler breeze than a fan. In hot dry areas the diurnal variations are much greater and may be as much as 25° C. in the desert. To suit these conditions, the building should have a thick roof and thick walls with a large thermal capacity, and should be ventilated thoroughly during the night, when the temperature is relatively low, and left unventilated by day.

In many hot countries the air is made to circulate through the houses by a judicious arrangement of the apertures to take advantage of any slight breeze. In the south-west of Madagascar, for example, a hot and airless region, the natives stagger the windows of their cabins. In the oldest Egyptian and Mesopotamian houses the doors were arranged in this way to encourage the creation of draughts. Holes were also made in the upper part of the walls and in the roof to let out the hot air, while fresh air was brought in from the north side through air-shafts similar to those used on steamers. The artificial ventilation of houses by means of air stacks has been practised since the very earliest times in all countries bordering on the Persian Gulf.

Ventilation towers of traditional antiquity are still in use in certain parts of Persia. They are square, brick structures rising about ten feet above the roofs of the houses. Each of their four faces is provided with apertures, generally arranged diagonally to capture the wind from whatever direction it blows and to direct it down into a room on the ground floor through a shaft fixed to an opening made in the middle or in a corner of the ceiling. The shaft is fitted with a shutter and when the latter is opened to create a draught, the occupants can stand in front of it to enjoy the fresh air. These ventilating stacks are seen particularly at Lindja, on the Persian Gulf, at Goum, at Yezd to the south of Isfahan, and at Seistan. A similar device is found at Hyderabad, in the valley of the Indus, where many of the houses are fitted with air-shafts, covered with a cowl of sheet-iron. Standing erect on the roofs, they have a most strange appearance.

External Protection

In some very windy countries, and especially on open plains, it is not always possible to find a fold in the terrain which will shield a house. Even if the house is low and squat without any protrusions, specially designed to stand up to gales and provided with an additional revetment on the most exposed front, it may still be necessary to have recourse to external protection. In this case, very thick hedges, or rows of trees planted close together, are generally used to make a screen.*

The Provençal farm is shielded from the mistral by a dense row of cypresses, which have thick foliage and will grow in compact groups without being affected by their neighbours. These trees flourish in the Mediterranean climate, and, being both flexible and resistant, they form an excellent protective screen.

The scattered farms on the very windy plateaux of the Caux country are completely hidden behind a wall of beech trees and elms, which protects the apple orchards at the same time. In very exposed districts in Normandy and Brittany, there is a tendency to replace the trees with cement walls, which are used in other places to screen sensitive fruit trees from the cold wind.

In the Eifel Mountains in Germany, the houses are also protected by hedges and a very interesting study has been made by Smeets of their use on the high plateaux of the Ardennes in Belgium. To soften the rigours of the climate, almost all the houses are surrounded with hedges, which vary in height but give the landscape of the whole region a distinctly individual aspect. Sometimes, as at Sourbrodt, Mont and Ovifat, the farms are scarcely visible, being hidden on all sides by hedges of beech pollards in which openings are made to serve as doors.

The tree most often used is the beech. Neither the wind nor the cold disturbs it† and it can live a hundred years. It has the advantage of keeping its leaves throughout the winter, when they effectively fill the interstices between the branches, and of shedding them in the spring, when the north wind is less keen and the weather is milder. The hornbeam is also used, not alone but in conjunction with the beech, because it grows quickly and is useful for filling an accidental gap. Its branches, however, do not become so intertwined as those of the beech.

---

*Translator's Note.—A compact row of trees suitably placed above houses or orchards situated on the slope of a hill can be used to protect the building from the effects of cold katabatic winds which sometimes roll down the slope during the night time.

†Translator's Note.—Except that it becomes distorted if it is planted in districts where icy winds are prevalent in the early spring when its shoots are appearing.

The spruce is less often used because there is the danger that its roots may spread under the house and it has the further disadvantage that it loses its lower branches rather early in its life. Nevertheless, screens consisting of conifers may be seen at Longfaye and at Francorchamps.

Smeets shows that the protective hedges on the plateaux of the Ardennes have changed in the course of time with the habitations that they shelter and that they are now tending in many places to disappear. The old rural thatch-covered cottage, which used to be small and low, was surrounded with a hedge, taller than itself, sometimes thirty feet high, planted thirteen to sixteen feet away. As the houses gradually became more solid, the practice of surrounding them completely with a screen of beeches was allowed to lapse and often only the hedges breaking the force of the winds from the west, north and north-east were retained, these winds being the most dangerous because they often damage the roofs. Moreover, the milder climate at lower altitudes has attracted some of the inhabitants of the Hautes Fagnes, who have gone to settle there permanently. As at Bévercé, near Malmédy, the hedges which they still plant round their new houses are lower and are intended only to give shelter from the prevailing wind. The old houses built of pisé are gradually disappearing and the construction of new houses continues to improve. Tiled or slate roofs have been substituted for thatch. The use of double-glazed windows is spreading and the walls are now protected by zinc or fibre plates. There is therefore less need for protective hedges, which are considered to be an obstruction to light and air and to absorb ground moisture. If one is already in existence on the site for a new house, it is generally left untended or kept simply for its artistic effect. Hedges are retained as screens only in certain districts, such as Sourbrodt, Elsenborn and Xhoffraix, where the wind attains exceptional force.

On the Hungarian Plain it is customary to plant trees round houses and stables, and in many countries outside Europe this precaution is taken where there are great plains with no obstacle to break the force of the wind. At Fargo, in North Dakota, for example, the farms stand in the middle of a thicket of trees. The Plains of Canterbury, in New Zealand, are covered with rows of eucalyptus and conifers and small fields are surrounded with hedges of broom. The low wooden farms lie concealed behind very thick, wide yew hedges to protect them from the *north-wester*, which comes down from the summit of the Southern Alps.

The Bataks of Sumatra, who live on a plateau where the monsoon blows with great force, have been obliged to build their houses in groups and surround them with earth banks or walls planted with trees, which are quite deformed by the action of the wind.

The Weathercock

This chapter would not be complete if it did not mention the weathercock, which, in open country, perched on the gable of an old family house and sensitive to all the caprices of the wind, is one of the patriarchal features of the building.

The smoke rising from a chimney is undoubtedly the best weathercock because it shows the slightest change in direction of the merest breeze. It is, however, a very old custom in many countries to place on the top of the roof a metal vane which revolves round a vertical axis and indicates the direction of the wind.

Little is known about the use of the weathercock before the tenth century. For a long time it had the character of a symbolic ornament, reserved by strict laws for ecclesiastical and manorial buildings. The weathercock on a church generally took the form of a cock in gilt metal; in civil architecture it represented the emblazoned banner that the nobility had the right to hoist on their turrets and the free towns on their belfries and it was then called an escutcheon. The armorial bearings were painted on it or cut out of sheet metal. At the end of the Middle Ages weathercocks were sometimes quite large and their use was very widespread from the fourteenth to the fifteenth century. Later they were cut tastefully or ingenuously in the most bizarre shapes and erected on the roofs of houses or outbuildings. They were often decorated in a simple and imaginative style with an object or a scene taken from everyday life. It became the custom in some places, especially in the district of Nevers, to use them as trade signs. They then showed a stranger as he approached a village the way to the inn or the blacksmith's shop. They can still be seen in the country today, but they are not now made in such picturesque shapes as they were formerly. Very often they are rusty and half worn out, creaking as they turn on the roof of some old deserted house. The weathercock, once a graceful attribute of old houses, has been relegated to the rank of a museum piece and no longer has its place in modern domestic architecture.*

*Translator's Note.—This is not really true in England. Quite a number of old weathercocks are carefully preserved and are often used in the modernization of old houses.

CHAPTER VIII

The Wind as a Transporting Agent

THE wind plays a considerable part as a medium of transport.
As we have seen in a preceding chapter, it drives along groups
of clouds, dispenses moisture and regulates the distribution of
rainfall. Its role, however, is not only to transport enormous
masses of water vapour; it also purifies the air of large towns
by carrying away the smoke and it promotes the growth of
vegetation by dispersing seeds. As soon as it acquires sufficient
speed, it is capable of holding solid particles raised from the
surface of the ground in suspension for several days and of
carrying them over long distances.

Considered from the human point of view, the action of
the wind sometimes appears beneficial, but it may often be
disastrous. In the majority of cases, while large-scale destruction is taking place, men can only stand by helplessly, like the
farmer in Kansas sadly watching his fields being turned into
dust by one of those terrible storms that periodically devastate
the Great American Plains. Some of these performances of the
wind are less spectacular than the awe-inspiring *black blizzards*
of North America, but they can be just as dangerous, when,
for instance, the wind spreads disease-bearing microbes and
causes epidemics.

Unable to struggle against the atmospheric currents, men
have tried to turn them, in certain circumstances, to their own
advantage. Spherical balloons make use of the wind as a means
of transport and during the First World War the combatants
resorted to it, when its direction was favourable, to spread
poison gas.

Deposits of Loess

The word *loess*, of German origin, comes from *lose*, meaning light or friable, and is a popular name given by the inhabitants of the Rhine Valley to a superficial deposit which is very common in that region and contributes largely to its fertility. Loess is a Quaternary formation, which is spread uniformly over the whole area. It is like soil in appearance, soft to the touch and pale yellow in colour, and is made up of fine particles of quartz, carbonate of lime and a small quantity of a clayey substance, which is coloured yellow by hydrous iron oxide and serves as binding material.

The origin of *loess*, which is widely spread over the earth's surface and is particularly favourable to cultivation, has caused much discussion among geographers and geologists, who are not even now in complete agreement on the subject. Some are of the opinion that it was brought down by the rivers, while others—and these seem to be in the majority—think that it was carried by the wind. Probably both factors contributed to its formation. The disagreement seems partly to arise from the fact that the word is used in too wide a sense and typical loess is often confused with silt, which is of alluvial origin.

Loess is usually confined to the temperate zones and, as it always contains stems of herbaceous plants, it is permissible to think that it must have been deposited where there was a type of vegetation like that found on the steppes.

A very well documented study made by Mademoiselle V. Malychef (Ref. 47) shows that loess in Europe and Asia occupies a large area stretching from the Atlantic almost to the shores of the Pacific. In Europe its southern limit passes through the forty-second parallel, but in Asia it extends below thirty-four degrees north. Keilhac estimates the area covered by loess on the Eurasian continent to be 6,250,000 square miles, which, with that in the Americas, makes a total of 10,156,250 square miles, although of course, this does not imply unbroken distribution.

In Europe, the loess deposits are divided into two large bands oriented east-west; the northern lies on the edge of the mountain masses of the west and centre of the continent, the southern coincides with the zone of Tertiary folds. The first band includes the south of England, the north of Brittany, the Paris Basin and the north of France, Belgium and Holland, a part of Germany, and Poland. The southern band extends over the basin of the Garonne, the middle basin of the Rhône, the Swiss plateau, Lombardy, the basin of the Rhine, the plain of Bohemia, and the valley of the Danube. These two bands join to the east in Russia to form a vast layer which covers the upper and middle basins of the Dnieper and the Volga, part of the basin of the Dniester and the steppes of the Crimea, the basin of the Don, the northern steppes of the Caucasus Mountains and the Caspian depression.

In Asia, loess ordinarily covers the plains and plateaux at the foot of the mountains or fills the low-lying areas between them. The zones are divided into two groups: the western comprises Syria, Palestine,* the Transcaucasus, Persia, Afghanistan, Turkestan and Siberia, where it extends over the steppes to the south of the Taïga as far as the Transbaikal. Tibet, Mongolia, Manchuria, and China belong to the eastern group. The Chinese loess, known by the name of "yellow earth" (Hwang-tu) is particularly developed in the valley of the Hwang-ho (Yellow River). Some authorities estimate that it can reach a depth of 1,300 to 1,950 feet, but Le P. Licent estimates it at 500 to 650 feet, and Anderson at only 160 to 200 feet.

In the southern hemisphere, the principal area of loess deposits corresponds to the Pampas in South America. There are also large deposits, known by the name of *cangahua,* at an altitude between 8,200 and 10,820 feet in the valley between the Andes in the north of Ecuador.

The true loess of north-east China, which is the best known, originated in a region having the characteristics of the steppe and most authorities agree in attributing it to the action of the wind, which is thought to have carried it from the Gobi Desert. A close study has been made of the loess deposits in Russia in Europe; some authorities ascribe them to floods caused by

---

Translator's Note.—An interesting paper on *The Aeolian Soils of the Northern Negev* was given by S. Ravikovitch, of the Agricultural Research Station, Rehovot, to the International Symposium on Desert Research held in Jerusalem in May, 1952. It has since been published in the Proceedings of the Symposium.

melting glaciers, others to the action of rivers in flood, while there are some who think that the wind played the principal part in their formation. At one time it was generally considered that the loess deposits in the Rhine basin were of alluvial origin, but Richthofen attributed them to the action of the wind and this now seems to be the prevailing opinion. Many geologists agree that the constituents of loess come from the Quaternary glacial deposits of northern Germany; others, however, think that they must have come from the silt of the Rhine floods.

The question of loess is worth consideration, for this kind of deposit, in whose origin the wind seems to have played a predominant part, is of direct concern to mankind because it is always favourable to cultivation and often of great fertility.

Transport of Volcanic Ash

Sometimes the ash thrown off by volcanic eruptions is transported over long distances by the wind and this is a phenomenon which may have some effect on the whole of the human race. The most important is that the particles of volcanic glass in suspension in the atmosphere absorb the solar radiation in direct proportion to their quantity; it is therefore not surprising that periods of cold are recorded after great eruptions, since the ash prevents a proportion, estimated as ten to twenty per cent, of the sun's heat from reaching the ground. This happened after the eruption of the Krakatoa in 1883, when the ash was thrown up to a height of nearly seventeen miles. Three years elapsed before it had all fallen to the earth again. During that time it was driven in all directions by aerial currents and when it reached Europe it slightly obscured the sky for nearly two years.

A considerable fall in temperature was also recorded in Argentina, particularly at Mendoza, after the eruption of the volcanoes in the Andes in April, 1932. Violent west winds blowing constantly in the same direction rapidly carried large quantities of ash as far as Buenos Aires, situated at a distance of 875 miles from the volcanoes in Chile. The ash formed a screen, blocking out the sunlight so that, in the middle of April, it was dark during the day in the Argentine capital. The streets and the housetops were covered with it and it made the fields in the Pampas dangerous for the herds.

Pompeii was buried (in A.D. 79) by the volcanic ash carried by a strong wind from Vesuvius. In 1906, ash from this volcano was collected as far away as Holstein and that emitted by volcanoes in Iceland has been carried by the wind to Norway and even to Sweden. In the New Hebrides, two volcanoes continually throw out large quantities—the Iahue at Tana and the Benbow at Ambrym. As a result of the régime of the southeast trade wind, it is nearly always carried in the same direction. In the immediate vicinity of the volcanoes, on the side down the prevailing wind, it forms a large and continually renewed accumulation which practically prevents the growth of any vegetation. Farther away, also to the leeward of these two volcanoes, the ash falling in smaller quantities enriches the soil because it decomposes rapidly and is favourable to growth.

The continual presence of these minute particles of glass in the air is, however, not without its disadvantages and it may be observed at Ambrym that a large number of Kanakas have bad eyes. In normal times the falling ash is not harmful, but during violent eruptions when it is extremely abundant, the trade wind carries it in such large quantities that it becomes a serious annoyance. It settles in thick layers on the houses, obstructs the gutters and water reservoirs and causes serious damage to plants.

The effect on agriculture of the transport of volcanic ash by the wind has been shown by H. Scaëtta in connexion with the volcanoes of Central Africa. Fine material, such as lapilli, sand and ash, thrown up into the air by volcanic eruptions, is separated according to its weight and ascending velocity and transported by the aerial currents of the different strata of the troposphere. Lapilli and ash are deposited not far from the centres of eruption; light lapilli are carried a certain distance by the trade winds and ash also seems to have been carried by the anti-trade wind of the southern hemisphere. Scaëtta says that the transport of this volcanic glass by the wind has had a very important effect, not so much because it has helped to form extensive new soils as because it has enriched the mineral reserves of various, and sometimes rather poor, soils.

Lapilli and ash are often absorbed into the texture of the
soil by the infiltration of water. The direction of the winds in
the upper strata of the atmosphere is also important in affecting
the composition of soils. Light volcanic particles, projected
more than 9,000 feet, are caught and carried westward by the
currents of the trade wind. The land over which these currents
pass, especially the Horst in the Congo, has been and still is
periodically renewed by falling ash. The land situated to the
east does not have the advantage of this supply of mineral
brought by the wind, which affects the chemical composition
of the soil all the more as it consists of basic glass which decomposes very rapidly and introduces new or fresh mineral elements according to the nature of the matrix.

The Transport of Snow

If there were no wind, the snow falling in the winter would
form an almost uniform layer on the surface of the ground. It
certainly does not do this, especially in very windy countries
where the winters are hard. The wind drives the snow suspended
in the air and deposits it to accumulate wherever it may be
sheltered from further disturbance. Large heaps are formed
round mountain chalets and isolated farms in the country,
blocking their doors and nearly burying them. Piles of snow
brought by the wind sometimes lie several feet deep on the
roads, obstruct the railway lines and cut off communications.
Hamlets and villages thus find themselves temporarily isolated
after snowstorms and trains may be kept at a standstill in open
country for hours and even days in succession.

Very often the wind not only drives the falling snow almost
horizontally, it causes the snow on the ground, especially if it
is powdery, to rise and swirl in eddies, and carries it away to
pile up in sheltered places exactly like fine sand. For this
reason, in many cold countries, plateaux and areas much
exposed to violent winter storms, are often almost free from
snow. The snow, swept along by the wind, accumulates behind
the folds in the terrain and fills all the depressions. The slightest
obstacle may cause the formation of snow dunes, similar to the
sand dunes made by the wind in other places. The powdery
snow heaped up in the mountains by blizzards may preserve

Spruce devoid of branches on their windward side, Ile de St. Pierre

Acacia tree permanently bent by the prevailing wind

Curtain of poplars to break the violence of the wind at Martigny, Rhone Valley

Windbreak of conifers protecting houses against the wind at Springfield, New Zealand

A plantation of Scotch fir partly overwhelmed by sand at Maviston sand-hills, near Elgin, Scotland. The fluting marks are due to the tops of buried trees

Sand dunes encroaching inland at Culbin Sands, Elgin, in spite of marram grass planted to check movement of sand

A screen of she-oaks protecting the Suez Canal against blown sand

its fineness and crystallization, but severe and persistent cold generally makes it heavy, and at high altitudes it causes dry avalanches, called "low-temperature" avalanches, which are very different from those engendered by certain hot winds like the *föhn*. As a general rule, the action of a dry and cold wind on a stretch of snow from which it has carried away as much as it could is to encrust it by hardening its surface and causing protuberances and undulations which are the despair of skiers.

In some countries the wind is the principal factor in the distribution of the snow. J. Westman, a member of the Swedish Scientific Expedition, showed that this is the case at Spitsbergen. A large part of the snow falling in winter on the low land of this archipelago is a result of the action of the wind. During the violent and very frequent storms between the south and the east, the wind sweeps up the snow lying on the mountains to a height of 1,300 to 1,650 feet, letting it fall again on the low land. It may also lift up part of the layer which has settled there and cause extremely violent eddies. Westman remarked that the hurricanes coming from the south partly rid the mountains on the north-west coast of Spitsbergen of their snows which are transported either to the low-lying areas, where they disappear in the summer, or to the neighbouring ice-bank which is driven to more southern latitudes in the spring debacle. The wind thus deprives the glaciers of part of their supply and consequently reduces their size, although, in other places in the archipelago, squalls may pile up enormous masses of snow in the valleys and in this way help locally to increase them.

In the Juras and the Alps in Central Europe, and in North America, palisades are erected to prevent the formation of snowdrifts on the roads and in the railway cuttings. These palisades, however, do not always afford sufficient protection for railway lines in the mountains, and in very exposed places in Norway wooden tunnels are constructed to prevent the snow from obstructing the line.*

* *Translator's Note.*—Such palisades and wooden tunnels are also used on the West Highland line between Crianlarich and Fort William, where the line crosses the bleak and desolate Rannochmuir.

G

Transport of Plants and Animals

The wind plays an important part in the dissemination of seeds, which is very useful in forestry. In Germany, and many other countries, tree felling is done against the wind, in bands of 980 feet, to encourage this natural reafforestation.

Swarms of insects are often carried by the wind, like the *orthoptera* which attack certain regions of Africa and other parts of the world and cause great damage to crops. The acridiens, which are incorrectly called "crickets", belong to several species. In Algeria, the most formidable are the Moroccan locust and the migratory locust. From Africa, clouds of locusts may be driven by the wind over the Atlantic and are occasionally seen in the Azores. Even in France, showers of locusts are not unknown and there was a serious invasion round Arles in 1613. The World Meteorological Organization has now appointed a meteorologist to study the influence of weather conditions (including winds) on the movements of swarms of locusts in Africa—*Trans.*

Showers of frogs, fish and cockchafers have also been seen and were at one time regarded as miracles. It was not then known that the wind was responsible for these phenomena, which still occur from time to time. On the morning of 1st June, 1938, a thick cloud of butterflies darkened the sky over the town of Hanoi. Within a few minutes the traffic was slowed down by a glutinous mass of them which covered the roads. The insects were a species of Lepidoptera (*Hannibalis ataxus*), which is very rare in Tongking and is normally found in the mountains of Luzon at an altitude of more than 5,570 feet. This extraordinary swarm of butterflies seems to have been driven along in a secondary cyclonic depression by a typhoon passing north of the Philippines.

It was also a violent wind during a thunderstorm which brought the strange shower of frogs over the little town of Alexandria, Ontario, on 2nd August, 1939. The streets were literally covered with them.

CHAPTER IX

Sand Storms

THE nature of the soil is a very important factor in the origin of *sand storms*, which do not occur where the ground is moist or where vegetation has become established. Very dry and loose soils are eminently favourable to this kind of phenomenon, which reaches its peak in the steppes and the deserts where the wind can raise an enormous quantity of sand and dust. Sand storms principally occur, therefore, in the hot and desert areas of the world and in the regions bordering on them, and the fine sand may often be carried a long way from its place of origin.

The inhabitants of the countries subject to sand storms have given them special names, many of which have passed into the current language, so that there is a large variety of terms for the same phenomenon.

Sirocco, Kamsin, Simoom, etc.

It is a customary saying that the *sirocco* is the breath of the Sahara. This burning, dust-laden wind, which blows from the desert to the shores of the Mediterranean, has a well-known enervating effect on men and animals. It can also be very harmful to agriculture, especially in the spring, when it scorches the crops.

The sirocco, which is produced when a depression passes over the western Mediterranean, is felt all over North Africa, where it causes a considerable rise in temperature and a reduction in the relative humidity. It is called the *chichili* in southern Algeria, the *chili* in Tunisia and the *ghibli* in Libya. The *ghibli*, unlike the sirocco in Algeria, is not abated by its passage over high plateaux, and, coming straight from the Sahara, it is particularly dangerous for crops unprotected by screens.

In Morocco, it is not simply a wind from the desert, but also a descending wind which travels from the summit of the Atlas Mountains. It acquires, therefore, as it approaches the coast, certain characteristics of the *föhn*. This explains its very high temperature and extreme dryness, which are not solely due to its Saharan origin and its particularly frequent occurrence in the middle of the summer, but to the thermodynamic effect to which it is submitted in crossing the Atlas Mountains. E. F. Gauthier says that the *föhn* effect of the sirocco strengthens the influence of the Sahara and that, on the maritime slope of the Atlas, it is not very strong. It is a hot, scorching and slow-moving mass of air which strikes the ground vertically or at a slight angle. In spite of its lack of force, it raises clouds of dust that a stronger, but horizontal, current of air would not raise. At the first signs of this depressing wind, the Europeans defend their houses by sealing them against the outer air.

The sirocco is also felt in Sicily and sometimes in southern Italy and southern Greece, but when it reaches European shores it has become charged with a certain amount of moisture in its passage over the sea. Occasionally it comes to Madeira with great force from the south-east and may last for a period ranging from a few hours to several days. The *leste*, as the islanders call it, gives them a sensation of oppression and discomfort, and often brings clouds of locusts as well as dust and fine sand.

In Egypt, the sirocco takes the name of *kamsin* because it is supposed to last fifty days, or, according to another explanation, fifty hours. It is also said that this name is derived from the fact that the *kamsin* blows only for a period of fifty days round about the time of the spring equinox. Its duration is, in fact, very variable and fluctuates between a few hours and two or three days.

When the Suez Canal was being planned, one of the principal objections raised by the opponents of the scheme was that the canal would soon be filled up by the enormous quantities of sand brought by the wind from the desert. Experience has shown that these fears were unfounded. The Company has planted rows of she-oaks along the African bank to check the onslaught of the sand, but devotes only a very small part of its annual revenue to dredging the canal.

The term *kamsin* is also used on the coast of Somaliland,* but *ouari* is more common among the natives to designate the sand storms which are very frequent in the hot season from May to September. They contribute largely to making the summers so oppressive at Djibouti, where, when the *kamsin* is blowing, temperatures of about 50° C. are recorded in the shade. The heavy, suffocating air is difficult to breathe and a fine, brown dust penetrates inside all the houses. These sand winds may blow regularly, or in gusts, at very variable speeds, sometimes reaching sixty-two miles an hour. The dust that they raise and transport considerably reduces visibility and remains suspended in the sky long after the wind has dropped, forming a yellowish veil over the desert.

L. Lapeyre,[1] who has been the Director of the Meteorological Station at Djibouti for many years, says that the *ouari* usually rises at about 1 p.m. and drops either at 8 p.m. or at midnight. It causes very rough seas in the Gulf of Tadjourah and makes navigation impossible. The natives are particularly afraid of it, for it blows so fiercely that it can dismast their barques. They have therefore carefully observed the warning signs given by the clouds and have established an empirical but certain method of forecasting it which never fails and enables them to reach shelter in time. An hour before the *kamsin* begins the tops of the Mounts Mabla become shrouded in stormy cumuli. The sea is absolutely calm and takes on a leaden hue; then whirlpools appear in the distance and it begins to swell and rumble. The wind rises so rapidly that a boatman would have no chance to furl the sail of his barque if he had not been warned and already found shelter along the coast.

Translator's Note.—In the Somaliland Protectorate the *kamsin* is called the *kharif*. It was largely this wind which was responsible for the movement of the seat of government from Berbera on the coast to Hargeisa; the latter is situated on the high inland plateau where the *kharif* is not felt so keenly.

[1] "*Les vents de sable à la Côte Française des Somalis.* Ann. Phys. du Globe de la France d'Outre-Mer, No. 11, Oct. 1935, p. 158–9.

In the Sahara itself, the desert wind is called the *cheheli*, which means "south wind", although it generally comes from the south-west. In the Sudan it is called either *dchaoui* or *haboub*. In Mauritania it blows from June to September and is called the *iniffi*.

The word *simoom*, which had its origin in Arabia or Asia Minor, where it is pronounced *samum*, has become the most general term to designate the scorching winds which rise suddenly in desert areas, blow violently and raise the sand, then die as quickly as they began. It is derived from the Arabic word *samm* meaning "to poison". Sometimes in the East *samyel* is also used, a hybrid word coming from the Arabic *samm* and the Turkish *yel* (wind).

At the end of the First World War the English troops suffered very much from the simoom in Mesopotamia, and Lieutenant Normand (Ref. 57) carried out some interesting research on the physico-physiological causes of its action on the human body. This action is not simply due to the heat and the dryness of the air in contact with the body. The speed of the wind, more than its actual temperature, seems to be the essential factor in the physiological phenomena observed. The calm days in the summer are the hottest in the desert regions of Mesopotamia, but the simoom is a suffocating wind and the air seems hotter when it is blowing than it does in periods of calm. Sensations of heat or cold are, in fact, related to the gains or losses of heat through the pores of the skin. A sensation of greater heat, when there is not, in fact, a real rise in temperature, is explained by the reduced loss of heat, at a high temperature, when the wind is violent. In this case, the human body could even make a net gain of heat sufficient to be fatal. It is here that perspiration plays a vital part. Lieutenant Normand takes, as an example, a man resting in the desert and sheltered from the rays of the sun. The weather is calm. He is hot and perspires automatically, which has the result of lowering the temperature of his skin. In a hot and very strong wind perspiration is no longer adequate, for the temperature of the skin, then of the body, gradually rises, and this may cause a fatal heat stroke.

In fact, when the wind speed is very high, the air temperature itself becomes important. The wind brings heat by convection and when it is scorching and blows at too high a speed the activity of the sweat glands is insufficient. Consequently, for any temperature of the air above the temperature of the blood, there must be, in a given wind, a critical temperature beyond which there is a net gain of heat in the human body. These conditions (high temperature and strong wind), which occur precisely with the simoom, explain why it often has a fatal effect on human beings.

On the steppes of Turkestan, the simoom is known by the name of *tebbad*. It is much feared by travellers and is said, no doubt with some exaggeration, to have buried more than one caravan beneath the sand.

In Persia and Afghanistan the famous "120-days wind", blowing from the north between June and September, moves enormous quantities of sand. Its speed is reputed to reach 125 miles per hour in the desert of Seistan, situated astride the south of Afghanistan, the east of Persia and the north-west of Baluchistan. It raises veritable whirlwinds of dust and is supposed to have killed camels and sometimes men.

Dust hurricanes also occur on the Russian steppes, and on the great plains of North America they are real calamities. The dust storms of western America have such repercussions on the economy of the regions affected that the next chapter is largely devoted to them.

The north winds which sweep the Pampas in Argentina, raising clouds of dust, behave like a true sirocco, so does the *hot wind* in South Australia. In January and February this brings the fiery breath of the centre of the continent to the south coast and causes extremely oppressive heat waves at Melbourne and Sydney. The Australians call these hot winds from the interior *bricklayers* because they cover the houses with enormous quantities of dust.

Characteristics of Sand Storms

L. Petitjean (Ref. 60), the author of an interesting study on *sand storms*, divides them, from the morphological point of view, into three distinct types. Some take the form of "sheets", others of "curtains" or "walls", and those of the third category have the appearance of "columns". It would not be appropriate here to discuss their origin at great length, but it is interesting to note the essential characteristics of a very widespread phenomenon of which man has to suffer the injurious effects.

The season of sand storms varies in different regions. As a general rule, they are most frequent in the summer and most violent in the hottest hours of the day. In the region of the Red Sea and the Persian Gulf, they are prevalent from May to October; in the Sudan, from October to March; in the northern Sahara, from March to June. When they occur in the spring, as they do in Syria, they may cause serious damage to young crops.

Ch. Combier (Ref. 23), who made a study of the sand storms in the Syrian Desert, draws some interesting conclusions on their internal constitution. In a large number of cases they seem to be connected with circular movements round a horizontal axis. It is not the force of the wind which is the primary cause of a sand storm, but the fact that it is a descending wind which rebounds on the ground and draws up the dust.

All who have seen the approach of a sand storm describe it as startling. "The air is abnormally calm, hot and heavy," Combier writes, "the horizon seems blocked by a yellowish mass which gradually increases in size as it approaches. Then it becomes a high opaque wall. Aviators encounter it well above 3,000 or even 6,000 feet. There are strong blasts of burning air, then again a sinister calm: the wall is quite near. Suddenly the wind rises; your skin is cut by fine particles of sand; they get into your eyes, ears, nose and mouth, and you are forced to find shelter. Indefatigably, sometimes for several hours, the wind snatches up all the dust that it can raise from the ground, until, in the most tightly shut houses, everything is covered with a thick layer of it."

The area affected by a sand storm varies considerably; sometimes it may be very large. Petitjean mentions a storm that occurred in the Sahara on 24th April, 1934, over an area several times larger than France. The wind may lift the sand up to 9,800 feet.

Dust columns, such as the *dust devils* in India, have much smaller proportions than sand storms in the form of "sheets" or of "walls". They are small whirlwinds round a vertical axis, similar to the dust eddies which occur in the summer in Europe. These columns rarely appear singly; they usually arrive in a series and are caused by the contact of two currents of air having a different speed or direction. They are very common in arid zones at any time of the year on days when the breeze is slight and the ground is heated by the sun's rays. They are nearly flat at the beginning, but rise rapidly to a column with a diameter of 60 to 100 feet and a height of from 300 to 600 feet, or even more. These *waltzing jinns*, as they are called in the Sahara, advance for a certain time, often in long files, in a concerted and very graceful movement, while at the top the grey or yellow dust forms a horizontal veil stretching over the plain.

Sometimes, with their rapid circular motion, they have sufficient suction effect to remove heavy objects and cause damage in their passage, but generally they are not destructive.

Very fine clouds of dust sometimes remain for a long time suspended in the atmosphere after the sand storms have passed and are driven along by aerial currents to form dry fogs that can be observed from the sea at some distance from the coasts. The celebrated Arab geographer, Edrisi, in the seventh century, spoke with astonishment, although he did not know the cause, of the reddish fogs which frequently obscured the sky over the Atlantic between the Cape Verde Islands and South America.* He called this the "Dark Sea" (Bar-el-Mecdolin); it was the "Gloomy Sea" of the authors of the Middle Ages.

Petitjean says with justification that sand storms make a very deep impression on the human mind by their grandiose character, their destructive power and their physiological effects. "With the hot breath of the simoom," he writes, "Nature suddenly seems to be on fire; the sky is covered with an immense copper-tinted veil. Suddenly a storm begins and drowns the scene in thicker and thicker mist. Gusts follow gusts. A fine dust penetrates the tiniest crevices, mixes with food and drink and gets into the eyes, nose and parched throat of the unfortunate traveller whose face starts bleeding as it is struck

Editor's Note.—It seems improbable that Edrisi would know of either the Cape Verde Islands or South America.

by dozens of small, sharp pebbles, thrown up by the wind from the rocks and dunes. Blinded, tortured with thirst, and unable to find any landmark by which to take his bearings, he advances more and more painfully until his forces finish by betraying him and he falls exhausted to the ground. The sand then buries him in its burning shroud until another storm comes to uncover his remains. How many bones of men and animals have been found by chance on the tracks of the desert! A legend has been handed down through the ages of the army of Cambyses engulfed in the shifting sands."

A sand storm is certainly an impressive spectacle and one which might well take hold of popular imagination and give rise to many legends. Nevertheless, not all the tragic stories, relating how entire caravans have been taken by surprise and utterly destroyed, can be accepted without reserve. Some authors, reliable in other respects, have described the ravages of sand storms in an arresting but certainly very exaggerated fashion. Buffon, for instance, whose work contains some very informative pages on the role of the wind, makes statements which leave us very sceptical. The following passage will serve as an example. "A very dangerous wind," he wrote, "that the local inhabitants call 'samyel', often rises during the summer along the Persian Gulf. It is suffocating and deadly, even hotter and more terrible than the wind in Egypt. Its action is like that of an eddy of boiling vapour and those who become enveloped in it cannot escape its effects. A wind stifling to men and animals also rises in the summer over the Red Sea and the lands of Arabia and carries such large quantities of sand that many people think that this sea will, in time, be filled up by the successive layers which fall into it."

But, when some allowance has been made for exaggeration, the fact remains that sand storms, if they last for some time, can be extremely dangerous and even fatal to travellers surprised in the desert.

In cultivated areas they are undoubtedly destructive. Moreover, as they greatly reduce visibility, which may become practically nil in the horizontal sense, they are very dangerous to aircraft flying over desert countries. M. Crouzet reported a sand storm of unprecedented violence which occurred at Kidal, in the Sudan, during the afternoon of 3rd June, 1933. For a quarter of an hour the town was plunged in dense fog and any object two inches away was completely invisible.

Sand storms also interfere with wireless communications because the grains of sand, as they strike against the objects which they encounter, generate electricity. These discharges make reception on the long wave in desert areas impossible while the storms last.

Mud Showers

Mud showers, which have always startled men by their unusual colour, are a phenomenon similar to sand storms and in the past they excited real terror. Petitjean gives some curious details on this subject. Homer, Virgil, and the chroniclers of the Middle Ages called them "showers of blood". Gregory of Tours relates that in the year 582 one of these showers so terrified the Parisians that they rent their garments, which were stained with what they believed to be blood.

The reddish colour is due simply to fine particles of sand transported over a long distance by winds from the desert. The particles carried in suspension turn the clouds yellow-ochre or brick-red and are dropped in a shower when the clouds encounter conditions of humidity and temperature which cause them to break up into rain. When the aerial currents charged with sand strike against a high chain of mountains, they are forced to ascend; the precipitations are then more abundant and snow takes on a yellow or red hue.

This phenomenon is comparatively frequent in North Africa and southern Europe. Two remarkable mud showers were recorded in March, 1901, and April, 1926. In 1901, the quantity of dust carried from the Sahara and falling in Europe was estimated at nearly 2,000,000 tons. This gives an idea of the part played by the aerial transport of dust in the formation of certain sedimentary deposits. Petitjean cites other memorable mud showers; one in the south of France in October, 1846, was so heavy that the mud obstructed the gutters and in December, 1859, an area of nearly 15,625 square miles in Germany was covered with reddish snow.

CHAPTER X

Soil Erosion by the Wind

SAND storms are primarily due to the extreme dryness of the climate and to the absence of a protective covering of vegetation on the land, both being conditions existing in deserts. Occasionally, however, they may occur in temperate zones where the rainfall is slight or irregular. In normally moist conditions, if the ground is covered with vegetation, erosion proceeds only very slowly. Here and there the wind may raise small quantities of soil and carry them from one place to another, thus contributing to the formation of new soils, but in districts where there are no trees and the vegetation is limited to short grass, the deterioration of the covering of vegetable growth may produce an extremely rapid acceleration of the process of erosion.

Man is obviously not responsible for the drought, nor for the fact that some soils are easily carried away by the wind, but he has largely helped to aggravate the dangerous effects of the climate by cultivating regions whose meteorological conditions made them unsuitable for permanent cultivation and by destroying the protective vegetable covering placed there by nature. This happened on the Great Plains of North America, where large tracts of land which ought never to have been cultivated intensively, were sown with corn. The real remedy for this condition—rain—is outside human control; the only effective measures that can be taken are to restore as far as possible the original character of the vegetation and to avoid repeating the mistake in the future.

It is not only in North America that soil erosion involves

considerable economic loss. It is equally menacing over large areas of Russia, Asia and Africa, and is very prevalent in the south of Morocco and in the valley of the Oued Mellah, near Casablanca, whenever the soil is denuded during the dry season. Almost everywhere measures are taken to check this danger by creating a suitable covering of vegetation; generally graminaceous plants, which grow quickly, are used for the purpose.

The Great Plains of North America

It is difficult for us in Europe to form an idea of the immense ravages caused by the wind on the Great Plains of the Middle West and in the three prairie provinces of Canada. It ruins the soil by raising whirlwinds of dust, with the result that vast areas which had been put under cultivation have been abandoned as completely sterile.

The Great Plains are characterized by low rainfall and very strong evaporation, by very hot summers with extreme differences of temperature and winds which are particularly frequent at the end of the winter and the beginning of the spring. They lie at an altitude of 2,950 to 3,900 feet, slightly inclined to the east; they are largely open with very few undulations and there is no obstacle to break the force of the wind, which can blow in all directions.

Several terms are used to designate dust storms: *dusters,* *dry blizzards* and *black blizzards* are the most common. The expression *dust bowl* is applied in the United States to the whole area where they occur and chiefly includes North Dakota and South Dakota, Nebraska, Kansas, Colorado, Oklahoma, New Mexico and Texas. In the west of Canada, the provinces most afflicted are Manitoba, Saskatchewan and Alberta, but not all areas suffer to the same extent.

Soil erosion by the wind is not a new phenomenon; it occurred on the Great Plains nearly every spring and occasionally in the autumn and winter. Violent dust storms had already been observed in the west of Canada in 1887 and they were mentioned more than fifty years ago in old reports on Kansas. Normally, they did not cause very serious damage and as soon as the young crop had begun to grow and hold the soil, after the first spring rains, the danger was over. After 1900, however, they became much worse. This effect seems to be directly connected with the development of stock-rearing and agriculture in the countries concerned. In 1927 and the following years, especially between 1932 and 1935, dust storms assumed truly catastrophic proportions as a result of particularly unfavourable climatic conditions. After a prolonged drought the crops were retarded in their growth by lack of rain, and the storms in the spring of 1934 were the worst ever known in the north-west of the United States and in the west of Canada since agriculture had become established there. At that time, the sky over the north of the plains was often darkened with clouds of dust, some of which, coming from the great corn-growing areas of Canada and from the States of Dakota, Nebraska and Minnesota, were driven by the wind as far as the Atlantic coast.

Both in Canada and in the United States, the struggle against the menace of wind erosion has become a matter of national importance.

Causes of Soil Erosion

This formidable phenomenon has several fundamental causes, which, although they appear simple, have not yet been fully explained. The essential feature is the action of the wind on a light, dry soil devoid of a covering of vegetation. Dust storms are particularly menacing during periods of drought or when hordes of insects have destroyed the vegetation and denuded the soil.

The farmers who took possession of the prairies after the American Civil War found them almost intact and they kept their enormous herds of cattle on these vast stretches of grassland. In 1890 they turned over to sheep. Continuous grazing soon deprived the grass of its protective character, leaving large patches of the land bare. The war in 1914-18 took thousands of farmers away from their farms and again favoured cattle-rearing on a large scale. The existing herds were increased to a million head. At the same time nearly 50,000,000 acres of land were ploughed and sown with wheat. After the ploughing, not a blade of grass was left. This agricultural development exactly coincided with the beginning of a period of exceptional humidity, which lasted for several years. The crops thus established conveniently helped to fix the soil. The farmers did not pay any attention to this fortunate chance and neglected the usual methods, which were not then so well known as they are now, of protecting the soil against wind erosion. Then, in 1930, the succession of abnormally wet years came to an end and was followed by the worst period of drought that has ever been recorded in the Middle West. The drought, combined with the lack of precautions and an insufficient covering of vegetation, caused soil erosion on a large scale. The corn land disappeared in clouds of dust and half of the cultivated area was soon turned into a desert.

Different soils are affected by erosion to very different degrees. Their resistance to it depends on their physical nature, that is, on the size and arrangement of the constituent particles. Those which have a uniform texture and a fine granular structure are the most affected. Nevertheless, neither coarse sandy soils nor heavy clay soils are immune. Those with the finest texture show the most resistance and can remain intact under cultivation for some years, but as the quantity of organic matter that they contain gradually diminishes they also disintegrate and are swept away.

Experience has shown that the simple action of strong winds on dry and denuded soils does not necessarily cause serious erosion. The fact that fine particles of dust are lifted high in the atmosphere by ascending currents indicates that turbulence also plays a part in it. Temperature too has an important influence. A sudden change in the direction of the wind, without any appreciable reduction of its speed but accompanied by a large drop in temperature, results in a marked diminution of erosion. A number of observations also showed that recently ploughed virgin soil was rarely swept away. As long as the fibrous particles from the previous vegetation remain, they give the soil cohesion. In the end this vegetable matter decomposes and erosion occurs at the first favourable opportunity.

The repeated action of frost greatly favours it. This fact has been observed many times in the south of Alberta, where dust storms are not unknown in the depths of winter. Periods of frost alternate with the hot chinook winds which rapidly melt the snow and dry out the upper layer of soil. Obviously, where the land is covered with snow, there is no risk of erosion in the winter. The danger comes particularly in April and May and, less frequently, in June. Finally, a very important factor in soil erosion is the method of cultivation employed.

Consequences of Dust Storms

The first result of dust hurricanes is that formerly fertile land is reduced to the condition of a desert because the useful top layer of soil is swept away. They also cause damage in districts which have not been affected by the erosion by burying cultivated areas under heaps of dust.

The action of the wind on the soil is rather like that of a sieve. It lifts the lightest and most fertile particles high in the atmosphere and deposits them perhaps hundreds of miles away. The coarsest and less fertile particles roll along the surface and pile up behind obstacles. Heaps of dust several hundreds of feet long and twenty feet high have been known to collect in the fields. When a dust storm occurs at the beginning of the spring, it carries away the seeds on the point of germination, thus destroying the future crop. If it arrives later, the speed of the wind laden with sand cuts the roots and branches of the young plants and again the crop is lost.

In Dakota and very exposed districts the ranches are generally protected by wooden palisades to prevent them from being covered by the light moving soil. Often the palisades themselves become completely buried and lost to view.

Soil erosion, in addition to the direct loss to agriculture, raises several other serious problems. It increases the cost of maintaining the railways, which have to be kept unencumbered. Roads also become buried and impassable and ditches are filled in. The dust attacks not only the farms themselves but it forces its way into the houses, however tightly they are shut, and penetrates even into the beds and the food. In Kansas, which may have forty dust hurricanes in a year, the inhabitants have to dig caves in which to take refuge. The dust, driven eastwards from the prairies, invades even towns and villages situated some distance away. Above the dust bowl the sky is inky black.

Lug-sail boat in the Bay of Aiguillon

A lateen-sail boat off the coast of S. India

Flettner rotor ship at New York

A modern glider taking off, towed by a motor vehicle

Crescent shaped wind-shelter made of mangroves near Khor-Anghar, Strait of Bab-el-Mandeb

Wind shelter in the form of a wall, made by desert nomads at Oued Boli (Somaliland)

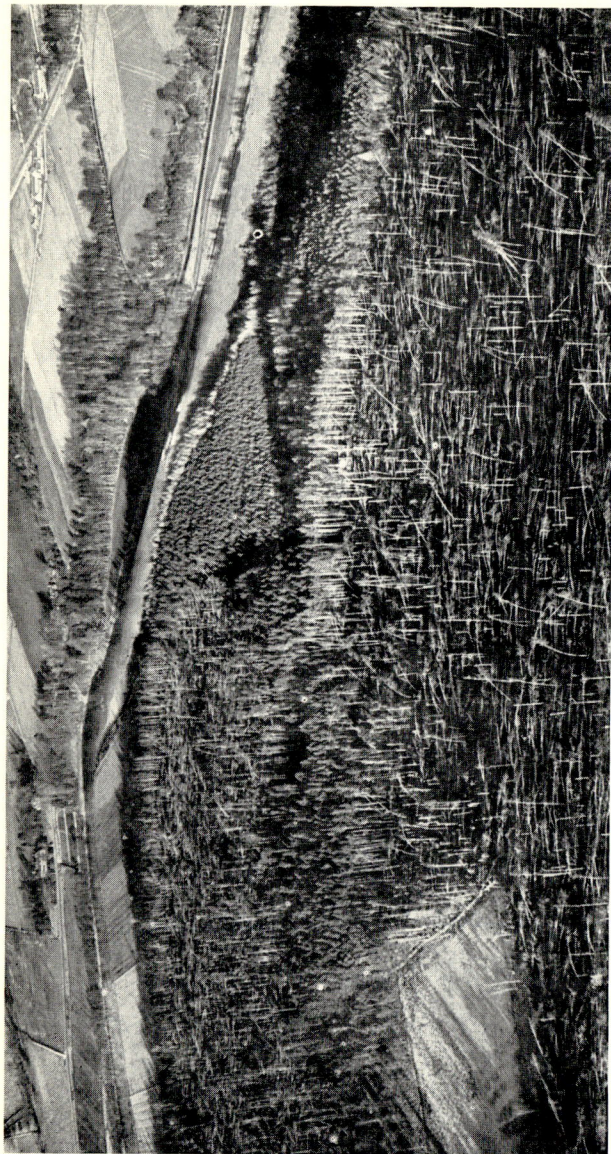

Damage caused by the great storm of January, 1953, in the conifer forest of Ballogie, Aberdeenshire

The dust is sometimes so thick that it causes motor accidents on the roads and in the streets of the towns, which have to be artificially lit in the complete obscurity. It brings on a disease which is called *dust pneumonia*.

The following figures forcibly show the extent of erosion on the prairies. The area affected by dust storms in the Middle West States is nearly 321,000,000 acres, a quarter of which is now practically useless for agriculture. One hundred and sixty-five thousand persons have deserted the prairies since 1930 and each new dust storm is the signal of departure for several thousand others.

Experiments made in 1934 at Regina, Saskatchewan, showed that the quantity of dust contained in a cubic mile of air, and carried by the wind in a storm, was about 200 tons. As some storms rage over several thousand acres, it can be realized how great the total displacement may be. In Nebraska and South Dakota, one of the great hurricanes of 1934 carried away a quantity of soil estimated at 500,000,000 tons. The soil may be affected to a depth of one foot and, in serious cases, two feet. It is a well-known fact that the effects of erosion are cumulative and that once the process has started each successive storm aggravates it. As the climate, to some extent, deteriorates at the same time as the soil, simple means of defence in a district are not generally sufficient to remedy the evil and, once a desert has been created, it becomes permanent.

Means of Combating Soil Erosion

Experience over the last few years has shown that wind erosion, if preventive measures are neglected, is capable of utterly destroying cultivated areas. Where the soil has already been degraded, the original covering of grass must be restored before a plan for the protection of future crops is put into operation.

Various methods of prevention are now applied on the Great Plains, but there is no special method which will be effective in all cases. Every farmer must decide what will best suit his own conditions.

A wise precaution is to plough in a grass crop, as vegetable matter helps to bind the particles of soil together. In areas

H

where dust storms are particularly serious, some farmers successfully practise the system called "strip cultivation". The land is divided into long, narrow, parallel strips, oriented perpendicularly to the direction of the prevailing wind. Alternate strips are cultivated and the others left fallow. As a precautionary measure, a light crop of oats, which does not reach maturity, is often sown on the fallow strip at the end of July. On the plains of Texas, maize or Indian millet is planted, either of which protects the soil very effectively. After the wheat harvest, the straw is left on the field to form a covering. In some places, ploughing machinery has been designed to raise small mounds of earth which serve as protection.

Certain crops are also cultivated with the young wheat until it has taken sufficient hold of the soil. The variety of protective crop depends on the district and on the type of soil. Beans, as well as Indian millet and maize, give excellent results. One method of protection frequently used consists in digging rows of deep furrows perpendicularly to the prevailing wind. At first sight, this might seem to favour the action of the wind, but actually it does not, since the wind is more dangerous blowing over a flat and uniform, than over a rugged, surface. These furrows hold back a large part of the dust and have the further advantage of conserving the rain.

The success of all efforts to combat erosion depends on concerted action. The precautions taken by one farmer will be useless if dust from neighbouring territories falls on his fields. It will promote the disintegration of other soils and the menace will spread rapidly. One single unprotected piece of land may cause erosion over a wide area.

CHAPTER XI

The Peril of Dunes

An important phenomenon caused by the wind is the formation of dunes, which are found in nearly all latitudes and all climates. In fact, the majority of dunes distributed over the earth's surface belong to two distinct zones: the sea-shore and the desert.

The morphological study of dunes belongs to the science of physical geography and it would not be appropriate here to discuss their origin, evolution and classification,* but the consequences of the movement of sand by the action of the wind relate to human geography because they are generally disastrous for man and force him to struggle against this threatened invasion.

In all sandy regions there is a local and often very rich nomenclature to designate the different types of dunes and this abundance of terms in itself shows how important the phenomenon is considered.

Coastal Dunes

Most coasts are covered with sand cast up by the waves in quantities that vary with the shape of the shore. A low and shallow beach is more sandy than a steep one sinking rapidly into the sea.

The rate at which the sand collects depends on its dryness,

---

*Translator's Note.—A book by Dr. R. A. Bagnold, F.R.S., *The Physics of Blown Sand and Desert Dunes*, Methuen & Co., 1941, deals with this subject. Dr. Bagnold also presented a paper, *The Surface Movement of Blown Sand in Relation to Meteorology*, to the International Symposium on Desert Research, held in Jerusalem, in May, 1952.

fineness and density and on the force and direction of the wind.
It is therefore essentially irregular. The condition most favourable to the formation of dunes occurs when the coast line is
almost perpendicular to the direction of the prevailing wind.
Often the banks of sand that have piled up along the shore
remain stationary for some time and are gradually fixed by
the vegetation which develops spontaneously on their surface.
When the sand has made great inroads and the wind is very
strong, the dunes, which vary greatly in depth, advance rapidly
inland.

The dunes in many sandy districts were much more stable
before they were inhabited than they are today. When the
covering of vegetation which had taken root on the surface
was destroyed to make way for crops and pastures, the solidi-
fication accomplished by nature was disturbed and the dunes
once more became mobile. In New Zealand, particularly, the
colonists, attracted by the vast stretches of grassland covering
some of the groups of stabilized dunes, considered them to be
excellent pasture for their sheep and they set fire to the thickets
and brushwood which seemed to them an encumbrance. The
animals then made gaps in the vegetation, which were further
enlarged by the wind. In many cases, therefore, the thoughtless
destruction of the vegetation which had become established
on the surface of the dunes increased their instability and turned
them again into enormous masses of moving sand.

Advance of the Dunes

From a mass of detailed information about the movement
of the dunes in the south-west of France the main interest lies
in the incontrovertible evidence that the dunes have moved
and are moving steadily eastwards.

The advance of dunes, besides being a threat to man's work
and his home, has other consequences which affect him. Sand,
driven by the wind, can appreciably modify the existing topography. In the Pampas in Argentina, the natural boundaries
separating individual holdings often consisted of sandbanks
which were displaced when the Pampero blew. Mobile dunes
can transform the contour of a coast; they can fill up ponds

and lakes and completely obstruct the mouth and change the course of a river.

Sandbanks enclosing the lagoons which are often found along low-lying coasts may be due to alluvial deposits but the wind can achieve the same result, especially in stormy weather, by raising waves which pile up enormous masses of sand along the shore. The sandbanks on the shore of the Bay of Biscay are an excellent example and many others of similar origin can be seen along the shores of the North Sea and the Baltic, as well as on the coasts of the Gulf of Guinea and of the United States and on the east coast of Madagascar.

Many ports which were prosperous in antiquity have been silted by the combined action of the wind and the waves. Such was the fate of the ports of Acre, Tyre, Tortosa, Tripoli, and Laodicea in the eastern Mediterranean, which were all obstructed by the sand brought by the enormous waves raised by the west winds.

The combined action of the wind and the sea, which has so often had the result of devastating the shores of islands and continents and of destroying land which man has brought into cultivation, may also have the opposite effect and give him, on occasion, new lands which he can turn to excellent account. For example, in the small archipelago off the coast of New-foundland, a sandbank, the Isthmus of Langlade, has joined the two islands of Langlade and Miquelon. The salt ponds on the Baltic coast, especially in Latvia, are being filled up by sand from the neighbouring dunes, and in Egypt the sands from the Libyan Desert are increasing the alluvium of the Nile.

The Sahara dunes, advancing under the influence of the north-east trade wind are gradually filling up Lake Chad because the tributaries flowing into it cannot compensate the loss from evaporation caused by this very dry wind for six months of the year. One of them, Bahr-el-Ghazal, has already had its course diverted by the dunes. The Niger formerly flowed into the Djouf depression, but as the sands have barred its northern course it has made a new bed along the belt of dunes until a tributary flowing into the Gulf of Guinea captures its waters.

The wind, which tends to heap up sand in depressions,

often fills up river-beds in desert areas. The rivers then flow underground and their course is indicated by the presence of oases. Wells can be sunk to bring the water to the surface.

Stabilizing the Dunes

The advance of dunes, under the influence of the wind, is a constant menace to human endeavour. If no attempt is made to stop it, large stretches of territory may become permanently barren. The stabilization of dunes, to achieve a positive result, must be a methodical enterprise bearing on the whole of the threatened area. It is therefore more successful when it is undertaken by the state than when it is left to the initiative of individuals. The most effective means to this end is provided by vegetation. There are, in fact, in all countries, certain plants whose fibrous roots are capable of retaining the sand and gradually fixing the dunes.

Brémontier succeeded in stabilizing the dunes on the coasts of Gascony by making large plantations of pines and, as a result of his work, similar efforts were made in other countries, as, for instance, in Germany along the Baltic coast, and in Holland and Belgium, where the coastal dunes are nearly forty-four miles long and one and a quarter miles wide. In Belgium they were planted with various resinous species of trees. The cluster pine, in particular, is used in places very much exposed to the sea winds while, further inland, the Austrian pine and the Corsican pine are preferred. Leafy varieties that offer very little resistance to the wind are used only for the *pannes* or *lettes*—the small valleys that separate the dunes.

In Denmark, the greater part of the dunes which formerly occupied a belt of 156 square miles in the west of Jutland and the north of Zealand have been stabilized with the help of graminaceous plants and pines. The moving or "white dunes" have become "grey" or dead dunes.

In Spain, to check the advance of the coastal dunes of Guardamar, the method of Brémontier has been adopted or sometimes that used to preserve the vines of the Ampurdan, threatened by sand from the dunes of Las Rosas, which consists of setting out lines of protective plants, at intervals of

six and a half feet, perpendicularly to the direction of the
prevailing wind and cut at right angles by other lines ten feet
apart. The stabilization of the sand is completed by spreading
earth over its surface in the proportion of 220 to 330 cubic feet
per acre.

An interesting study of P. Buffault (Ref. 13) describes the
efforts that were made in New Zealand to stop the progress of
the coastal dunes.* Before 1908 it was largely unheeded; the
dunes were considered to be a natural defence against the sea,
which frequently encroached on low-lying coasts like those to
the west of Wellington and in the neighbourhood of Dunedin.
But as they advanced towards the interior and began to cover
valuable land, they became a serious menace and the struggle
against them a question of national importance. At the present
time, they occupy about 470 square miles, especially between
Cape Maria van Diemen and Wellington in North Island
and from Cape Farewell to beyond Westport in South Island,
as well as on other parts of the coast and even in the centre
of the country, where inland dunes are found round Rangipo
and in the district of Otago.

The objects proposed in the fixation of the New Zealand
dunes were to protect the coast against the encroachments of
the sea and to prevent the sand from obstructing the mouths
of the rivers. For this work, the methods of Brémontier in
Gascony were adopted and particularly those used in Germany
along the Baltic. In addition there is, in New Zealand, a large
variety of plants which live in sand and which, in many cases,
are capable of spontaneously fixing it without human inter-
vention.

The stabilization of dunes usually has a two-fold object:
first, to check the progress of the sand and then to bring it
into cultivation by establishing useful plants on it.

The experience acquired in Europe as well as in the United
States, in Australia and in the Cape Province, has shown that
it is preferable, before planting conifers, which, when they are
young, are in danger of being uprooted or buried by the action

*Translator's Note.—An article on the coastal formation of New Zealand
entitled *Accidents and Interruptions in the Cycle of Marine Erosion*, by C. A. Cotton, was
published in *The Geographical Journal*, Vol. CXVII, Part 3, September, 1951.

of the wind, to fix the dunes with herbaceous vegetation. There are several plants suitable for this purpose and the species naturally vary in different countries. Beach-grass (*Ammophila arenaria*) has given good service and is widely used to check the progress of the dunes in northern France near Malo-les-Bains. It grows spontaneously in Europe and North America and is propagated in three ways: by sowing, by means of its numerous and creeping roots which spread about thirty feet and may be buried six or ten feet in the sand, and finally by pricking out the young plants. It is the last method which is the most commonly used for stabilizing the dunes systematically over large areas.

The work of stabilization always starts from the origin of the deposits. When this is the sea the plantations proceed from the shore to the interior. It is the usual custom to make a first dune of defence with stakes and branches and to protect it immediately with beach-grass.

Inland Dunes

In arid countries where the wind is absolute master the dunes are more extensive than in maritime regions. The scarcity of rain and the extreme dryness of, the air does not allow them to become stabilized by natural means, like many coastal dunes, unless a notable change of climate occurs later and an increase in humidity, marking the recession of the desert, favours the establishment of vegetation and transforms the sand which was formerly mobile into "dead dunes". A change of this kind has been observed in certain places in the Niger Colony.

Contrary to general belief, the sand zones are small in relation to the total area of the deserts, which represent about 7 per cent of the continents of the world. What is usually called a "desert" is more often rocky and barren land, or even steppe, than a sea of sand. There are, certainly, vast areas covered with sand and dunes in the Sahara, the Kalahari, Arabia, Baluchistan, Turkestan, the Gobi and in Central Australia, but they represent only one aspect of the desert. It is estimated, in fact, that dunes occupy only one-ninth of the Sahara. The largest groups are found in the Iguidi, the Djouf, the Erg, the

Edeyen and in the Libyan Desert, which undoubtedly has the greatest mass of dunes in the world.

To the south of Constantine, which is visited by so many tourists, the dunes mark the beginning of the desert at the foot of the Atlas. Thus L. Aufrère (Ref. 3) is right in saying that "it is not surprising that the Sahara is sometimes represented as an immense desert of sand and the sand as the very essence of the desert".

It is easier to oppose the invasion of coastal dunes which threaten to engulf villages and cultivated land than to combat the progress of inland dunes, especially in desert areas, where, except on the edge and in oases, it is impossible to grow protective plants. In most desert countries in the world, traces of ancient cities, dating from an era when the climate was more humid, have been found under a thick mantle of sand and dust. The excavations undertaken in India, in Mesopotamia and in the Syrian Desert have thus brought to light cities which were flourishing in antiquity and which, in the course of the centuries, became covered with sand. The dunes of the Libyan Desert caused a number of ancient Egyptian villages to disappear. Today many settlements in arid zones are threatened with the same fate, as, for instance, In-Salah, in southern Algeria, which is encircled by dunes. Near Kashgar, in Chinese Turkestan, there are some villages, situated in the midst of cultivated areas, which are being slowly buried by the advance of the sand.

All over Fezzan the natives surround their gardens with matting to protect them from the dunes which are threatening the oases. The inhabitants of the oases of the Souf in the eastern Erg display great energy and tenacity in maintaining gardens and date plantations in the midst of dunes. Their cultivations have a rather singular appearance, for they occupy more or less circular depressions between the dunes (*siouf*) which intersect in all directions. The depressions, deepened to reach a more humid level, are given the name of *haoud*.

"The ridges of the dunes," writes M. Aufrère, "serve as field boundaries which tend to become displaced or demolished by the wind. But there are men who are experts in the knowledge of the wind and who bear the name of *riaha* or 'men of the wind'. By moving the palm hedges the wind experts of the Souf succeed in orienting the action of the wind and in combating the progress of the sand. But outside the oasis, the wind does what it will." At the bottom of the hollows among the dunes about ten palm trees may be growing, sometimes less. In spite of all the care that the Soafas take in raising their palm fences, the dry sand is so mobile that the lightest breeze makes it fall into the bottom of the *haoud*. Their gardens would soon be buried if they did not work unremittingly to remove the sand to the top of the bank.

In the centre of Australia, the invasion of the sand is a constant threat to the crops and to the stock-rearing farms, which in those arid regions are supplied with water by means of wind-driven pumps. Several have already been covered with sand and have had to be abandoned. Elsewhere the progress of the sand has considerably reduced the extent of the stock-rearing territory, and farmers who raised 100,000 sheep thirty years ago consider themselves fortunate to be able to maintain 40,000. Others who formerly had 6,000 cannot now keep more than 2,000.

CHAPTER XII

The Wind as a Source of Energy

"Moving air," said Jean Brunhes, "is a natural force of considerable importance to man. The wind was of valuable assistance to him at a time when he was less spoilt than he is today." Certainly, of all the natural forces, it is the one of which he made the earliest use—often, it must be admitted, very inefficiently—long before he thought of using the energy from watercourses and tides or the steam from geysers. For centuries the two great inventions for taking advantage of the wind were the sailing ship and the windmill, but the graceful sailing ships that formerly ploughed the seas and the mills that ground the corn and enlivened the countryside with their gay aspect have nearly disappeared. Today, man almost completely rejects the services of the wind, asking no more of this inexhaustible source of energy which nature has put gratuitously at his disposal than that it should pump a few cubic feet of water and supply a few incandescent lamps in rural districts, when it could generate thousands of kilowatt-hours.

The wind, according to A. B. Lamb of Harvard, is capable of supplying in a year three-quarters of all the energy that could be obtained from the three conventional sources (coal, petroleum and gas) in 2,000 years. Another American scientist estimates that the wind distributes an average of seven horse power on each acre of the earth's surface.

There has, however, recently been a renewed interest in the wind as a source of energy; experiments are in progress and ingenious projects have been put forward with the object of making use of it in a much more rational way than in the past. The Danes, who started working along these lines and had made remarkable efforts to achieve a practical utilization of wind power, find that they can effect an economy of 60 per cent on the use of fuel with the help of wind turbines and windmills.

OLD WINDMILLS

Windmills appear to be extremely old and were known in the East in very early times. The Egyptians were already using them about 3600 B.C. to grind corn and pump water for the irrigation of arid land. They did not, apparently, make their appearance in Europe until the end of the Carolingian period, but they spread rapidly and were not long in replacing the hand mills which had been used up to that time. They were adopted in a large number of maritime countries where there was plenty of wind and for centuries their gay and lively silhouette and the humming of their machinery were a feature of the hills and plateaux bordering on the sea.

The classic windmill consists of a certain number of sails or blades, generally made of canvas stretched over a wooden frame, evenly arranged in a star formation round a horizontal axis and oriented perpendicularly to the direction from which the breeze comes. The majority of windmills have four sails; sometimes they have five or six; eight or ten are rather exceptional.

The upper part to which the sails are fixed usually consists of a revolving cap which can be oriented by hand to bring the sails into the wind. In some windmills the cap is equipped with a long wooden tailpole to make this operation easier, while others are worked automatically by a fan tail.

The shape of the sails and the main body of the windmill, as well as the materials used to make it, vary enormously in different places. Every country has its own type, or types, of windmill. France, in particular, has a large variety.

A much more unusual, but very simple, kind of mill is one with a vertical axis. This is a real wind turbine, but as it has a rather low efficiency—distinctly less than the classic type—it is not very widely used. Vertical axis mills with four rectangular blades of plaited straw are used principally in the district of Seistan and Baluchistan for grinding corn. I had the oppor-

tunity of seeing a unique mill of this kind in the Province of
Quebec in Canada. The classic types were fairly common
there during the French occupation and the remains of some
of them can still be seen round Quebec and Montreal. This
vertical axis mill is near Mont Joli in the hills overlooking the
St. Lawrence. It has twelve wooden blades each shaped like a
trough and is used by a farmer for pumping water.

The ordinary function of windmills is to grind corn, but
they are also used for oil pressing,* as they were formerly in the
district of Lille, for fulling in the textile industry, for draining
the land, as in Holland and Flanders, and in some cases also
for irrigation. Up to the end of the nineteenth century, many
Norwegian sailing ships built of wood were equipped with a
small windmill to pump out the water from the hold as ordinary
pumps did not always suffice to drain it.

Geographical Distribution

Holland very naturally comes to mind as being pre-eminently the chosen home of windmills, but many other countries
in Europe also had them, principally Spain, Portugal, France,
Belgium, England, Germany, and Denmark. Even in Iceland
they were used for grinding rye, but they have all disappeared
now. At one time, up to 900 were counted on the Åland Islands.

They were found sometimes spaced at intervals, sometimes
in series along the ridges of hills, and sometimes in compact
groups on slightly rising ground. There were thus real "colonies"
of mills, like the Knoll of Cassel in the north of France. Between
this district and Lille they were very numerous and part of the
town used to be called "Lille Mills". This name dates from the
time when Lille, with its colza mills, many of which were still
in existence a hundred years ago, was a real oil capital.

Most of the English† windmills were in Essex, Sussex,
Surrey and Kent and belonged to three main types. The *post*

Translator's Note.—A mill is used for pressing vegetable oil from the seeds of
certain plants such as linseed and rape.

†*Translator's Note.*—See *Windmills in England*, by Rex Wailes. The Architectural
Press, London, 1948. A detailed description of a post mill in Norfolk is given in
The Story of Sprowston Mill, by Wing Commander H. C. Harrison. Phoenix House,
London, 1949. Both these books contain a glossary of technical terms used in wind-
mill structure. *British Windmills and Watermills*, by C. P. Skilton, was published in
the "Britian in Pictures" series in 1947.

mill, so called because it was mounted on a post, was rect-angular in shape and made of wood. It was oriented into the wind by hand with the aid of a tailpole. The *tower mill* was circular and built of brick or stone. Only the conical upper part could turn so as to bring the sails into the wind. The third type, the *smock mill*, was somewhat similar to the second, but its hexagonal or octagonal tower was made of wood and rested on a brick base.

Today most of the mills which are still working in England are provided with a motor as a stand-by when there is no wind. Very few are used for grinding corn, their main function being to grind meal for the animals. When they are old and need costly repairs, they are usually allowed to fall into disuse.

Windmills still play a very important part in Holland. Most of them do the work of pumping. On the polders—lands which have been conquered from the sea or brought into cultivation by draining the marshes—they keep the water at a constant level corresponding to that of the collecting dikes. Groups of twenty, thirty or forty are generally placed in rows on slight eminences overlooking the pastures, especially round the great polders of Schermer, Purmer, Meemster and Wormer in the province of Noord Holland. Other much smaller mills, having a more modest task, can also be seen in most of the low-lying districts of the country. The farmers instal them in their meadows to pump out the water and drain it into the neigh-bouring canals. It is estimated that if the windmills and dikes were not maintained 38 per cent of the area of Holland would be inundated.

In the Swiss Juras, where water power is scarce, windmills were formerly used and one is still working at Mont-la-Ville (Canton of Vaud). Several were used to drive saw-mills, espe-cially at Gaicht, in the Bernese Juras to the north of St. Sulpice. In 1909, one of these old "wind saws" near Charbonnières, not far from the Lake of Taillères in the Vaudois Juras, was restored. The Island of Bornholm is famous for its saw-mills worked by the wind. Most of the large number of windmills on the Åland Islands were intended for grinding corn, but a few were also used for driving saw-mills, like those in the Swiss Juras.

The old windmills, whose existence seems to be coming to

an end in most parts of Europe, still survive on many islands, particularly in the Mediterranean where modern machinery has not yet been introduced. Jean Brunhes has given us a picturesque description of the mills on Majorca, white towers with a small thatched cone at the top, which are generally installed in series in the vicinity of the towns for which they grind the corn into flour. Rows of them may be seen near Palma, forming the "Mill Quarter" (El Molinar). They crown the hills round Selva, Inca, la Puebla and Muro and encircle the town of Manacor. They usually have six wings arranged round an axis carrying, perpendicularly to the plane of the wings, a large mast to which is attached a network of ropes radiating from the centre, which, seen from the front, looks like a spider's web. Each wing has its sail, rolled up when the mill is stopped and spread out when it is started by means of the ropes, which enable it to be suitably oriented according to the force and direction of the wind at the time of operation. "It seems," wrote Jean Brunhes, "that the landsmen, having seen clearly and at close quarters the use that could be made of the wind by shreds of sail attached to the masts of barques, have quite naturally perfected their mills by giving them a real set of sails governed by a spar."

The sails of the windmills on Mikonos and in the Greek Archipelago are very different from those used in France and Holland and are much more like the type seen in the Balearic Islands. The cap, covered with thatch and placed at the top of a white tower, is fixed and the sails are set to face the prevailing wind. They are arranged in the form of a propeller consisting of about ten wings, each carrying a small triangular sail controlled by means of ropes. Mikonos has for a long time been famous in the Ægean Sea for the number of its mills which have earned it the nickname of "Windmill Island".

Windmills still flourish in the Azores, especially at Fayal, where they are used to grind maize. The upper part of the mill turns on its stone base so that its powerful arms can be oriented in the most favourable direction.

The Decline of Windmills

After fulfilling for centuries their modest but very useful

role, windmills are on the point of disappearing. In many countries where they once brightened the landscape with their great sails outlined against the horizon, they have been still for several decades. The development of machinery, the electrification of rural districts and the competition of the large flour-milling companies have dealt them a fatal blow. Some of those which still survive are no longer used and many others, neglected for several years, are falling into decay.

In France, the millers, discouraged in the unequal struggle against the large flour-milling companies and crushed by taxation and the restrictions of the new agricultural legislation, have for the most part abandoned their mills. Only 5 per cent of those still working are used for producing flour; the others grind different kinds of grain for feeding livestock. As an increasing number of farmers now possess a small motor to grind the food for their animals on the farm, the mills are idle and their owners prefer to discard them rather than to carry out costly repairs to keep them in order. In Holland, although windmills are being used to drain the polders, many which used to grind corn and can still be seen dotted all over the country have practically stopped turning.

Protection of Old Windmills

Once they have been abandoned, windmills soon fall into decay. In many countries during the last few years, societies have been formed, not to set them working again, but to preserve and restore those which still exist so that the special character which they give to the landscape shall not be lost. In France, H. Webster (Ref. 74) has led this movement and is trying, with the help of the Society of the Friends of Old Mills, created in 1928, to save the last of the mills in the north near Lille, Calais and Dunkerque, along the Loire between Saumur and St. Nazaire, in the region of Beauce, in Morbihan and Finistère, and in Provence near Arles, where the legendary mill of Alphonse Daudet still stands. Those in Vendée and Charente are now no more than pitiful ruins and almost everywhere there are old windmills, stripped of their sails, which are used only as barns and week-end cottages.

The recent survey, made by the Departmental Commission

Tropical vegetation flourishing under the damp south-east trade
winds (New Hebrides)

A wind-driven saw at Landrienne (Canada)

Flemish windmill at Cassel

Vertical axis windmill for pumping water at
Quebec

Catavento wind-driven pump for sea water,
Brazil

Enfield-Andreau 8 kW wind-power plant working on the depression principle. The blades, turned by the wind, are hollow and throw out at their tips air which induces a current up the shaft. This up-current drives a turbo-generator at the base of the shaft of Sites, of the Moulin de la Galette or Blutefin at Montmartre (1939), which dates from the seventeenth century and has kept its mechanism intact, recalls how important windmills were for a long time in the Paris area. The exact date of their first appearance in the capital is rather uncertain. In any case, it goes back several hundred years, for, if one accepts the evidence of R. Courville (Ref. 25), the two things that most struck Le Tasse when he came to Paris in 1570 were the stained glass windows in Notre Dame and the windmills at La Butte (Montmartre), which were, even at that time, comparatively numerous. There were more during the seventeenth and eighteenth centuries and twenty-five could still be counted at the time of the Revolution. Afterwards they gradually fell into disuse and decay.

It is difficult today to locate the sites of these old windmills, although some of them are recalled by the names of the streets. Since the disappearance in 1917 of the Moulin d'Amour in Rue Ernest-Cresson, there is only one windmill in Paris—apart from the Moulin de la Galette—which is still recognizable as a windmill although only its tower remains. This is the Moulin de la Charité in Montparnasse Cemetery, built by the monks of Saint-Jean-de-Dieu at the beginning of the eighteenth century.

The towers of the windmills in the suburbs were often used as places of refreshment to which Parisians resorted on Sundays to drink light red wine and eat *galette*, a kind of flat cake made with flour, butter and eggs. They were the origin of the gay Montmartre of today.

A. Mousset (Ref. 56), in an interesting article on the old Paris windmills, recalls a curious incident which shows that a large number still existed in France a hundred years ago. An inquiry was started in 1836 to track down a secret organization whose object was to establish communication between Paris and Bordeaux by means of signals transmitted by windmills. This optical telegraphy began to function on 27th June, 1836. Coming two days after the outrage of Alibaud and at a time when France was at war with Spain, it greatly alarmed the authorities. Was it a new political conspiracy or a movement in favour of Don Carlos? The inquiry revealed that the sole

I

object of these mysterious communications was to convey to some speculators at Bordeaux "the current price of shares on the Paris Bourse in order to facilitate dishonest operations on the Bordeaux Exchange. We know," Mousset adds humorously, "that windmills have played a part in military espionage, but who would have thought that they would be accomplices in gambling on the Stock Exchange?"

A movement to protect the last of the old windmills has developed in Belgium and in Holland the society "De Hollandche Molen", founded in 1923, pursues the same object. A similar effort is being made in England and the Windmill Section, a branch of the Society for the Protection of Ancient Buildings,* is trying to save the beautiful mills of Sussex, Surrey and Kent and is making a methodical classification, county by county, of all the windmills in the Kingdom.

Even in the United States, a number of disused windmills are carefully preserved and, in spite of the great advance of mechanization in that country, two enormous windmills of the Dutch type can be seen in San Francisco pumping water for spraying the plants in the parks.

MODERN WINDMILLS

The old-fashioned windmills were in their time remarkably successful, but they had various disadvantages. They demanded a fairly strong wind, but not too violent. For this reason they could not work on an average more than two or three days a week and they needed continual supervision to supply what was on the whole not a very large amount of power. G. Lacroix observes that, while the water-wheel throughout the ages was undergoing progressive improvements, leading to the creation of modern turbines, the construction of windmills did not receive the same attention and for a long time made no appreciable advance. The four-bladed Dutch mills of large diameter (sixty-five to ninety-eight feet) remained practically unchanged for several hundred years.

*Translator's Note.—This society has issued a pamphlet on *Windmills* to which James Laver and Rex Wailes have contributed.

The Wind Wheel

The Americans certainly made a step forward when they designed the wind wheel with a number of large blades almost entirely covering its surface. Its diameter is usually twenty-three to twenty-six feet, although occasionally it may be as much as forty-nine. This invention has to some extent brought the wind back into use. Wind wheels, which are, unfortunately, not very pleasing in appearance, are mounted on metal pylons, and sometimes on the roofs of houses, and have in many places replaced the picturesque windmills of former days. They have also been introduced in districts where the latter were unknown, especially for pumping and irrigation in arid areas.

This type of mill performs valuable services in the country districts of the United States and Canada and its use has developed in most very windy places. Farms and isolated premises are supplied by the wind with electricity for lighting and water pumping, as well as for use with different kinds of domestic equipment like vacuum cleaners, washing machines, cream separators, etc.

Windmills play an important part in the irrigation of arid areas by providing power for bringing the subterranean water to the surface. It is in this field particularly that the greatest services may be expected from them. In the hot and dry north-west of Queensland in Australia there are several being used to pump water for the gardens from the gravel of the river beds. Today windmills serving the same purpose cover the great plains of the United States and Argentina and the plateaux in the Union of South Africa. Some can also be seen in North Africa.

The number of wind pumps has increased in the south-east of France and in many European countries, especially in the neighbourhood of Dresden in Saxony, where they pump water for the market gardens.

P. Deffontaines mentions the special use made of wind motors at the salt works of Cabo Frio, sixty miles north-east of Rio de Janeiro, from which fossil salt is extracted at certain points by saturating the soil with water. The salt water is pumped by means of Archimedes screws driven by wind motors. The mine is worked in the summer, although that is the rainy season, because the heat and the north-east trade winds which blow constantly from November to March facilitate evaporation.

The wind wheel giving a supply of electricity ought to be of valuable service in industry, but the energy generated by this kind of apparatus is at present very far short of what would be required for industrial purposes.

New Methods of Capturing Wind Energy

Steam and electricity have almost superseded the old windmills, except in cases where some irregularity in operation is unimportant. One of these is water pumping, either for drainage or for irrigation. It might have been thought that in the face of competition from these new sources of energy the manufacturers of windmills would have tried to improve them, but, having only a very rudimentary knowledge of the action of the wind on the blades, they could do no more than carry out a few simple modifications. The invention of the wind wheel obviously marked a real advance, but although it has been in use for nearly a century it has developed very little during that time and its application remains limited since it cannot be used in industry.

In a very interesting study on wind motors, G. Lacroix (Ref. 43) * recalls the aerodynamic researches made at Askov by the Danish professor, P. La Cour, from 1891 until his death in 1908. The results, which were applicable only to this type of mill, were, however, insufficient to form the basis of a general theory. Lacroix shows that, until recent years, the construction of windmills rested essentially on empirical rules few of which would be worth retaining today and which, in any case, scarcely constituted a theory. The manufacturers, having no scientific basis for their work, had to rely on practical experience until a fundamental theory was given to them indirectly as a result of the aerodynamic researches stimulated by the birth and development of aviation. It is largely due to the work

---

* *Translator's Note.*—Lacroix has written a very comprehensive article on wind power entitled *L'énergie du vent*, which was published in *La Technique Moderne* for 1st and 15th March, 1949, and 1st and 15th April, 1949.

of Prandtl at the Laboratory of Gottingen* during the First
World War that aircraft designers now possess a scientific
theory of the action of the wind.

After the First World War there was therefore renewed
interest in windmills. At the World Power Conference, held in
London in 1924, two papers were presented by Danish dele-
gates on the use of wind power. In the following year, tests on
modern windmills were made at the experimental station of the
University of Oxford. Two French engineers, L. Constantin
(Ref. 24) and G. Darrieus, independently of each other, designed and constructed light and very simple machines which
were successfully put into operation. The Darrieus wind-driven
generators, like most modern windmills, turn at a very high
specific speed and for this the blades have to be narrow, especially at the tip. Modern windmills differ in this respect from the
old type with their much larger blade areas. The small area of
the blades enables the apparatus, either when it is shut down or
in operation, to face the strongest winds without the need for
an effacing device. In many countries, inventors have thought
of adapting the results of recent aerodynamic research to the
utilization of wind power. The new windmills, with their
streamlined and clean-cut appearance, are therefore totally
different from their predecessors and also from the American
wind wheel. "To convert the kinetic energy of the wind into
usuable mechanical energy," Lacroix says, "the most appropriate equipment, as the work of M. Lapresle at the Eiffel
Laboratory has clearly shown, is the two-bladed wind turbine."
This turbine is a propeller made of hard wood; the blades
may vary in number, but are usually limited to two, and are
very similar in shape to aeroplane wings. The propeller drives
a dynamo, which charges the accumulators—hence the name
of "windcharger" given to this type of apparatus in America—
and enables the current generated to be used continuously. A
tail vane keeps the apparatus turned into wind and an automatic
regulator ensures its steady performance in a low or gusty

*Translator's Note.—A. Betz also worked on the aerodynamics of windmills
at the Laboratory of Gottingen and published an article, Die Windmühlen im Lichte
neuerer Forschung, Die Naturwissenschaften, Vol. XV, No. 46, 18th November, 1927,
and a book entitled, Windenergie und ihre Ausnutzung durch Windmühlen, Vandenhoek
und Rupprecht, Gottingen, 1946.

wind and prevents the blades from turning at excessive speed when the wind becomes too violent.

Many manufacturers are making special efforts to design wind-driven generators to take advantage of stronger winds, of a speed of at least 22 m.p.h. Without under-estimating the possible value of installations made in exceptionally favourable wind conditions, Lacroix rightly observes that modern wind-mills cannot become very extensively used unless the manufacturers deliberately set out to capture the energy from the low winds which are the most frequent in our part of the world. In France, the annual average wind speed is, in fact, only 9 to 11 m.p.h.* It is obvious, therefore, that the design of a wind-mill, unless it is intended for a very favourable site, should be calculated on the basis of an average speed of that order. In the United States, where the manufacturers are trying to popularize "free electricity from the wind", they recommend the use of windmills of 6 to 22 volts, starting to generate in a wind of 7 to 8 m.p.h. They must be mounted high enough to capture as much wind as possible, either on the roofs of houses, or, better still, on pylons in open spaces where there is no obstruction within a radius of at least 500 feet.

Today there appear to be some entirely satisfactory types of windmills; they are still only modest machines unsuitable for use on a large scale, but, as simple autonomous installations, they give good service in country districts which have not yet been given a mains supply of electricity.

Ambitious Projects

The wind is a capricious and irregular force, but the fact that it is free and inexhaustible could not fail to attract enterprising manufacturers in different countries, who have, in recent years, elaborated ambitious projects for capturing this enormous source of energy on a large scale with the aid of unusually large wind turbines. In most cases, however, the results obtained from wind-driven generators designed to operate in connexion with the network have been disappointing.

Translator's Note.—The annual average wind speed in Great Britain varies from about 17.5 m.p.h. in the Outer Hebrides to less than 10 m.p.h. in some inland districts of England and Scotland.

MacMaster in Australia and Darche in Algeria seem to have been the first to design modern large-power windmills. One project, which was never realized, was to build for the Chicago Exhibition in 1933 a tower 1,970 feet high, equipped with wind wheels capable of generating 100,000,000 kilowatt-hours annually. The generator placed on the top of this tower would then have been high enough to make full use of the wind whose speed is always lower near the ground.

An aeronautical testing station has been set up at Mont Ventoux in Provence at an altitude of 6,260 feet, where, as its name implies, strong winds are of common occurrence. At Carpentras, at an altitude of 328 feet, the annual record shows forty-three days when the wind speed is higher than 11 m.p.h. in contrast with 178 days at the summit of Mont Ventoux. This site therefore seems very advantageous for the utilization of wind power and it has been suggested that large wind turbines should be installed there. In fact, when it has been proved that such turbines can be used economically, there will be no lack of sites—principally on the coast—which will be favourable to the installation of wind power stations supplied by an indefinite number of wind turbines for converting the force of the wind into electrical energy. In 1933, H. de Varigny estimated that a series of 300 wind wheels, erected at suitable points on the Atlantic Coast, would be capable of satisfying the total demand for electrical energy in France and could supply annually 15,000,000,000 kilowatt-hours, for which it would be easy to find new uses.

H. de Varigny added that, if the installation of batteries of large windmills is envisaged, it would be useful to carry out a preliminary and methodical survey of atmospheric conditions in order to determine the best sites.* The wind is never con-

* *Translator's Note.*—After the war wind surveys were started in England by the British Electrical and Allied Industries Research Association and in France by Electricité de France. The two organizations have exchanged measuring instruments and collaborated in the collection of data. Two reports describing the methods used in this country were published by the Electrical Research Association, *Large-Scale Generation of Electricity by Wind Power—Preliminary Report,* by E. W. Golding, 1949, and *The Selection and Characteristics of Wind Power Sites,* by E. W. Golding and A. H. Stodhart, 1952. Two articles published in France describe the survey there, *L'énergie éolienne: sa valeur et la prospection des sites,* by P. Ailleret. Revue générale de l'Electricité, Vol. LV, March 1946, and *La recherche des sites qui pourraient convenir à une utilisation de l'énergie du vent,* by P. Ailleret. La Météorologie, April-June 1948.

stant, but in every country, mainly in coastal or mountainous districts, places can certainly be found where it blows more frequently and with more force than elsewhere.

In the opinion of those who envisage its use on a large scale, there is a great future for the wind as a source of energy. The inventors of the various projects consider that the quantity of energy that can be captured from the wind is practically unlimited and that its cost will be extremely low, comparable with that of the energy obtained at present from the most favoured hydro-electric power stations. If this expectation is realized, it will be an event of such economic importance that it is impossible to foresee all its consequences. These are obviously very optimistic views. Lacroix hopes that the research in progress will soon result in the erection of large-scale wind-driven generators for supplying electricity in rural districts, but he recognizes that in practice there are great difficulties to be overcome. "Although this motive power is free and at first sight seems very attractive," he says, "the installations necessary to capture and use it on a large scale have proved up to the present too costly to make its exploitation economical. It is possible—and, of course, it is to be hoped—that these difficulties will disappear in the near future in the light of modern technical research."

CHAPTER XIII

Sailing Ships

THE wind has given its greatest service of all to navigation. No doubt often inconstant and dangerous, it was for centuries the great propulsive force for ships at sea. The discovery of the sail enabling it to be used was made in the remote past, more probably by chance than by calculation. Vidal de la Blache (Ref. 72), in his *Principles of Human Geography*, has clearly shown the capital importance of this fortuitous discovery. "The use of the mechanical force of the air by means of the sail to overcome the resistance of the water contained the immeasurable germ of all future progress. It cannot be said of this invention that it had a character of universality. Many races of people who lived in contact with the sea did not know it at all or knew it only later. But it conferred an early superiority on those who, independently of one another, discovered its use. It marked them out. It forged peoples."

The Dawn of Navigation

A tree trunk drifting on the water no doubt inspired primitive tribes with the idea of the first raft and the first canoe. The invention of the oar and the paddle probably marked the beginning of navigation, but it is thought that the sail arrived later. The anchor, if it did not exist already, would inevitably follow. When the motion was produced by the oar alone, the crew, necessarily more numerous, could stop the vessel or haul it up on land. But the sail enabled the tonnage to be increased and the number of men to be reduced so that it became necessary to perfect the anchor as a means of keeping

137

the ship completely stationary at sea. Long after the invention of the sail steering continued to be done with one or two large-bladed oars. Median steering dates only from the thirteenth century when the oars or tillers began to be hinged on the stern-post by means of thongs and then of iron fittings.

The art of navigation remained for a long time rudimentary and it did not evolve in the same way in every country. Certain peoples, like the Phoenicians, the Greeks and the Arabs, showed a more progressive spirit than others; they realized the possibilities of this means of transport for trade, since it was simpler and quicker than long, expensive and dangerous journeys by caravan. It took several centuries, however, before navigation assumed a really scientific character.

Navigation on the High Seas

For coastal navigation the physical nature of the coasts and the wind régime were the deciding factors, certainly as far as the Mediterranean was concerned, but perhaps not for ocean navigation, as we shall see in connexion with the monsoons of the Indian Ocean and the voyages of discovery in the Atlantic.

"The difficulty," as E. Jurien de la Gravière (Ref. 40) picturesquely said, "was not to spread the wing of Icarus, but to dare to spread it in the open sea." The art of steering in the open sea when the shore is out of sight is certainly one of the most arduous problems that have ever confronted the human intelligence. It is therefore not surprising that it took centuries to solve it. Until the fifteenth century European sailors were mostly satisfied to hug the coasts, afraid of being driven out to sea. Sometimes, against their will, they were carried away by a storm. This happened in 1337 when a gale took an English barque to Madeira as, four centuries earlier, the Vikings had been taken to Greenland and North America.

The hazards of the wind have thus been the cause of a large number of fortuitous discoveries. The trade winds led Columbus to the discovery of America and in a similar way the Portuguese navigator, Gonzalo Velho Cabral, landed in Brazil. Even today, Japanese sailing ships may be driven by a storm as far as the coasts of the Philippines and some have been taken to Kamchatka. Canoes manned by the natives of Fiji are sometimes carried to the New Hebrides. Many of the South Sea Islands have become inhabited through the vagaries of the wind; for instance, a party of natives from Uvea in the Wallis Group, having drifted to one of the Loyalty Islands, settled there and gave it the name of their original island. But not all of these accidental landings on islands or continents, occasioned by the wind, are worthy of a place in the history of navigation. "What constitutes progress and is a real conquest is new land found by people who will know how to return from it and will have the means of going back to it" (E. Jurien de la Gravière).

Ships setting out in the North Atlantic and the Mediterranean encountered their greatest difficulties at the beginning of the voyage because they were then in the zone of variable winds and frequent storms. While the Polynesians knew how to take their course across the immensity of the Pacific, where the regularity of the trade winds so greatly assisted their progress, the sailors in the Mediterranean handled their sails on a sea where the winds are subject, especially during the winter months, to sudden changes without warning so that it was very often difficult to foresee where their capricious breath would drive the vessel.

Once they had passed the zone of variable winds, the navigators of the fifteenth century met, as they advanced south in the Atlantic, remarkably equable and constant breezes at the same time as a very mild temperature. They had entered the zone of the trade wind. It was, however, a long time before they passed the meridian of Cape Nun in the south of Morocco. In the era before the great voyages of discovery many sailors thought that, beyond the regions subject to known winds from which they were certain of being able to return, there were zones of untold danger which seemed forbidden to them. A terrible death was predicted for those who attempted to weather Cape Nun, where a number of Portuguese ships had been sunk in the strong currents. In spite of these fears, a party of Portuguese ventured as far as Cape Bojador with the encouragement of the Prince Henry known as the Navigator, although he himself did very little voyaging. In 1486, Barthelemy Diaz, more daring and driven by a violent storm, reached and passed the

Cape of Good Hope. He observed that the coast rose to the north and he gave the southern tip of Africa the name of the Cape of Storms. A little later, on 14th November, 1497, Vasco da Gama rounded this cape while on his way to explore the route to India.

Sails and Sailing Ships

The material employed for making sails varied very much in different places. Stretched skins were formerly widely used. The Veneti and the ancient Celts used leather sails, the Ona of Tierra del Fuego llama skins and the Alakalouf seal skins. The Melanesians made use of palm-leaf matting and the Polynesians of tapa. The Phoenicians and the Greeks had linen sails. Those of the Chinese junks are still made of plaited bamboo canes. Whatever might be the material provided by local conditions, the use of the sail enabled man to harness a natural force and put it to his service.

At the beginning, ships had only a single mast and one sail, but from the first century A.D. they had a supplementary square sail placed above the principal sail. The tonnage gradually increased. The Vikings were the first to have their hulls really protected against the sea and to practise the art of tacking, which enabled them to sail against the wind. Sails were progressively modified and perfected as the vessels improved. In the tenth century, the large ships of western Europe, raised at the ends, carried two or three masts. During the Crusades, navigation developed considerably in the Mediterranean, but the boats remained slow and sailed on a wind badly. The substitution of the triangular lateen sail for the quadrangular primitive type was a notable step forward. In the fourteenth century, the Genoese introduced a more mobile sail which later enabled ships to venture on the ocean and to launch out on great voyages of discovery. In the course of the following centuries, the evolution of the hull and the sail led to the elegant and rapid clippers of 1850 and to the enormous modern sailing ships which have only recently disappeared.

Sails are oriented according to the direction of the route in relation to the wind. The simplest case is that of a ship sailing before it. Sailors prefer the wind abeam, which has the advantage of reducing the roll. With skilful handling of the sails, they can then move faster than the wind. When it passes from the direction called "wind on the beam" and blows across the bow, it is still possible to make headway by orienting the sails as near as possible to the axis of the ship but in such a way that they still receive the breeze. The movement of the air tends to push the ship sideways so that it moves neither backward nor forward, but this leeway is checked by the resistance of the water against the hull and a certain useful effect remains which makes the ship advance in a direction inclined to the direction of the wind and against it. But when the ship is directly opposed to the wind, it can no longer move forward. The oblique direction that it can take against the wind depends on its nautical qualities. Racing yachts have to be able to sail on the wind, that is to say, as near as possible against it. They can hug the wind to the point of making headway at an angle of forty-five degrees to its direction and against it.

The Age of the Clippers

Towards the end of the eighteenth century during the War of Independence, the Americans designed the first clippers to run the English blockade. The town of Baltimore was justly famous for the construction of these sharp-built ships, low on the water and very swift. Later the word "clipper" was extended to include the large and magnificent trading vessels which were unrivalled between 1840 and 1880 and were the fastest that ever ploughed the seas.

At that time sailing ships, in construction, speed and beauty, reached their highest point of perfection. It would scarcely seem possible to achieve a more harmonious and beautifully balanced design than that of the clippers in their prime. Later attempts to improve and enlarge them were less successful. The enormous five, six and seven-masted vessels with steel hulls, exceeding several thousand tons, which were launched at the beginning of the twentieth century, were imposing and powerful but they no longer had the grace and elegance of the old clippers with wooden hulls.

No steamships of that day could equal the clippers in speed.

After their disappearance, it was about fifteen years before the progressive improvements made to engines and boilers enabled the fastest liners and warships to beat the speed records hitherto held by the clippers.

Clippers played a large part in the tea trade and in the illicit traffic of opium. At the beginning of the nineteenth century, the import of opium into China was forbidden, but it was smuggled along the coasts after it had been prepared in India. Although the risks were great, the profits were high. The opium clippers were small and easily handled ships of 250 to 400 tons. They could easily sail against the monsoon of the China Sea and escape pirates. "Towards 1860," writes L. Lacroix (Ref. 44), "the steamships captured the opium smuggling trade, but the tea trade offered some compensation to the clippers. The consumption of tea in the world continually increased. Freshly grown tea in full flavour was particularly in demand and the first consignments of the new crop arriving in England fetched high prices." The tea clippers took, on an average, 100 days to make the voyage from Shanghai to London, although some could accomplish it in ninety days.

Lacroix mentions one British ship-owner, James Bain, who had so much confidence in his ships that he signed a contract with the government in which he undertook to carry the mail from the United Kingdom to Australia in sixty-five days and to pay a heavy penalty for every day's delay. This contract was rigorously fulfilled.

Large Modern Sailing Ships

"The nineteenth century," wrote R. Chavériat (Ref. 20) in his beautiful book on *The Sail*, "was truly the century of sailing and it ended in the achievement of the most beautiful fleets of sailing vessels that have ever ploughed the seas." At the very moment when the use of steam was becoming general and changing all the old ideas of navigation, sailing ships, on the eve of their disappearance, became larger than they had ever been and attained a degree of perfection difficult to surpass.

After 1848, the tonnage was considerably increased and

the Americans built ships of 2,000 and 3,000 tons for the transport of emigrants to the west. At the time of the gold rush in
California the sea route from New York to San Francisco
round Cape Horn took about 100 days and was safer and
quicker than the journey over land through the United States,
where there was the danger of meeting hostile Indian tribes
and the difficulty in some places of finding water.

From 1850, the construction of iron hulls and later, after
1882, the construction of sheet-steel hulls enabled modern
sailing ships to be made very much larger. The average tonnage in the middle of the nineteenth century was scarcely more
than 400 to 600 tons. "Metal construction," Lacroix says,
"facilitated the development of new forms more suitable for
speed, and, in particular, enabled the hulls to be made much
longer in proportion to their width. This led to profound modifications in the rigging. Until then three-masted ships had been
considered the queens of the sea. With the straight hulls it was
necessary to reduce the height of the masts and to increase the
length of the yards. The age of four-masted vessels began. It
must be admitted, however, that before this period, at the
beginning of the nineteenth century, there had been four-
masted vessels with wooden hulls; some caravels formerly rigged
four masts."

With their slender lines, their steel masts, which were more
robust than those of wood and enabled the sails to be kept
unfurled even in strong winds, and finally the perfected form
of their sails, the great modern sailing vessels were able to
travel at twelve, fourteen, and even fifteen knots. Before the
days of the clippers, which had been specially designed for
speed but had only a small tonnage, a speed of over ten knots
was exceptional.

The need for large tonnage was felt towards 1890. The sailing ship was an economic means of transport, obviously suitable
for long voyages and for carrying heavy goods, such as wood,
corn, nitrates, and minerals, when speed was not important.
For this purpose the steamship could not compete with it. A
sailing ship can, in fact, carry a heavier freight than a steam
cargo boat of equal dimensions because it does not have to
carry dead weight in water, coal and machinery.

"In every port one could see those splendid vessels which for more than a quarter of a century were to fly their colours on distant seas" (R. Chavériat). In 1889 appeared the first five-masted ship, *La France*, 374 feet long and 49 feet wide. She was able to spread 6,579 square yards of sail and displaced 6,000 tons. For some time, before she was surpassed by German ships of still greater tonnage, she was the largest sailing ship afloat. The record was, however, to return to France a little later, when, in 1911, a ship-owner of Rouen launched the *France II*, of 8,000 tons. (She was lost in 1922 in the regions of New Caledonia.) From 1890 to 1914, hundreds of large sailing ships of all nations crossed the oceans, carrying to the most distant ports the heavy freights of which they had the almost exclusive monopoly. They took coal from England and loaded up with rice in China, jute in India, wood in Oregon, nitrates in Chile, wheat and wool in Australia and nickel in New Caledonia.

The long-distance sailing vessels on the route to Australia bore south across the Indian Ocean towards the fiftieth parallel to take advantage of the great west winds and often passed in sight of the Crozet and the Kerguelen Islands. Those which went to Chile and to ports on the west coast of the United States had to weather the formidable Cape Horn, which they rounded between the latitudes of fifty-six degrees and fifty-nine degrees south. There the perpetual westerly gales raise waves fifty to sixty feet high with a wave length of 1,000 feet from crest to crest, which make the tour of the globe without encountering any obstacle. The crew also had to contend with the cold, blizzards, ice-floes, and fogs. Those who have not sailed in the Southern Ocean cannot realize the fury of the squalls which fall so suddenly on ships, doubling or trebling the violence of the wind during storms. "They are accompanied by a raucous and deafening sound like the roar of a wild beast, which has earned them the name of *snorters*. Under the crushing force of these squalls, the waves of the sea seem to be levelled, then the moving glaucous hills are crowned with a fringe of boiling foam that old sailors call *greybeards*" (L. Lacroix). The large, heavily laden sailing ships met terribly rough seas in passing Cape Horn and during squalls it was very difficult to take in the immense sails with the ropes stiffened by the cold. The calms, too, though generally of short duration, were dangerous on account of the swell. The risks were further increased by the number of ships which passed the Cape. Those sailing before the wind to return to Europe cut across the path of those which were at the Cape or were tacking in order to make their slow and laborious progress westwards. As a result there were some disastrous collisions in which two ships sometimes disappeared. There were occasions when ships, unable to round Cape Horn after weeks of struggle, were obliged to make a detour by the Cape of Good Hope to reach the North Pacific. The French sometimes preferred to take this much longer but much safer route and to pass by Tasmania on their way to load up with wheat in California.

The time taken on voyages was very variable. At the end of the nineteenth century, the following were regarded as record speeds: 62 days from Europe to Chile, 72 from Liverpool to Sydney, 80 from France to New Caledonia, 79 from New Zealand to Europe, 109 from Europe to San Francisco, and 106 from Europe to Hong-Kong. In 1907, the *Félix-Faure* covered 1,000 miles in three days and even achieved 356 miles in 24 hours—an average of nearly 15 knots. Many voyages, however, took almost double the time—sometimes 153 days from Europe to Chile, 146 from Le Havre to Noumea and 191 from England to New Zealand.

R. Chavériat very justly remarks that these fleets of large sailing vessels helped to unify the world. "The great sailing ships," he said, "opened up communications, established the means of trade which have since developed to such an enormous extent and traced the regular routes of today."

The war of 1914-18 marked the end of the age of the great sailing vessels, which have progressively been abandoned for various reasons. In England it was difficult to recruit the necessarily large crews to man them. As more minerals are being treated in their country of origin, there is less heavy freight to be carried. The opening of the Panama Canal enabled the large modern cargo boats to compete with sailing ships on the west coast of the two Americas. In France, the eight-hour day, the lack of freight for export and the increase in taxation and

K

insurance have forced the owners of long-distance sailing ships to put them out of commission one after the other. The canal of La Martinière near Nantes, which once had the most beautiful fleet, became their cemetery after the First World War. The first to be laid up was the *Boieldieu* on 16th May, 1921. She was soon followed by twenty others. The last ended their sailing days in 1931.

The Finns have bought up the remains of the old French and English fleets and the home port of the last great sailing vessels is now Marienham in the archipelago of Åland. Of the thirty-one which were still sailing in 1935, twenty-six were Finnish. "But," writes Lacroix, "they disappear one by one every year and soon the great sailing ships with their square outline will be no more than a memory."

Navigation on Lakes and Rivers

Man probably used the sail on inland waters before he dared to venture on the sea. Rivers are particularly suitable for it, provided that they flow against the prevailing wind. In Brazil, the boatmen of the middle Sao-Francisco take advantage of the *vento geral*, which is a very regular easterly wind identical with the trade wind, to sail up the river. One has only to see the long lateen yards of the dahabiehs on the Nile, which was already being navigated perhaps as early as 6000 B.C., to realize that the prevailing wind in Egypt comes from the north. Similarly the barges with sails sixty-five feet high, which used to go up the Loire between Nantes and Orleans, showed the predominance of the west and south-west winds. The sailing boats on the Loire disappeared about sixty years ago but they were very numerous in the past. In the time of the Gauls this river was the chief route between the Mediterranean region and Britain. Its rapid waters and the long straight parts of its very slightly winding valley where the river current and the sea breeze exercise below Orleans their propulsive force in the opposite direction made it, as R. Dion (Ref. 29) shows, an eminently suitable river for rapid transport and distant communications. Going up the river, the loaded barges could cover the 197 miles from Nantes to Orleans in less than eight days when the wind was favourable and the level of the water sufficiently high. In the seventeenth and eighteenth centuries, light boats, called *cabanes*, carried passengers. In contrast with the Loire, the Seine, a slow-moving and very winding river, was much less favourable to sailing.

Many rivers had their own type of craft. The Garonne had the *filadière*, Holland the *weyschuit*, England the *wherry*. The boatmen on the Ganges have the *kistie* and the Chinese go up their rivers in sampans and junks. Lakes also have their own boats, which vary considerably from one country to another. There was a great contrast, for instance, between the Peruvian and Bolivian reed boats on Lake Titicaca, with a single rectangular sail, and the large boats on the Lake of Geneva, with two masts and lateen sails. These decked boats, ninety-two feet long, could carry 100 to 180 tons. There were still about a hundred of them in 1900, but they no longer exist today. Neither do the *cochères*, used for the transport of goods on the Lake of Geneva, which were small flat-bottomed boats, undecked except for a cabin, and also carrying two masts, each with a lateen sail.

The Decline of Sailing

Although there are still some sailing vessels today, it is often said that the art of sailing, which goes back so far in history, is dying because its use is becoming more and more limited and the time is approaching when it will no longer have any reason to live. Some people, however, refuse to believe that the decline of the sail is inevitable. Claude Farrère, in particular, declares, "The sail will not die as long as men demand their bread and their pleasure from the sea. No, the sail will not die. There will always be sailors because there will always be fishermen, yachtsmen and men who love the waves, the wind and the sound and swell of the sea."

Whatever the future of the sail may be, many maritime countries rightly consider that it provides an excellent discipline for all those who aspire to become navigators. They receive their instruction and complete their training on board sailing ships specially built for this purpose. Times have changed, but the great fundamental laws of navigation remain unaltered and the wind, even if it is seldom used now for the propulsion

of ships, still exists and is still master of the sea, which acknow-
ledges no other and rises and swells only in accordance with its
force. Everyone who ventures on the sea depends ultimately
on the wind.

The Magnus effect and the Flettner rotor ship

Gustav Magnus, a professor of physics at Berlin, discovered
in 1852 that when a smooth rotating cylinder is placed in a
current of air it is thrust sideways across the direction of the
current with considerable force. For more than half a century the
Magnus effect remained no more than a scientific curiosity,
but in 1923 Anton Flettner successfully sailed a three-foot
model boat employing Magnus's principle. This was followed
by the experimental ship *Buckan,* which was fitted with two
Flettner "rotors", each resembling a tall funnel capped by a
flat metal plate.

Each rotor was driven by an 8 h.p. electric motor, current
being supplied from a 45 h.p. diesel generating-plant. Stormy
passages across the North Sea and the Atlantic were successfully made. The advantages claimed for the rotor ship were
that the projected area receiving the force of the wind is only
about one-tenth that of a set of sails of equal power, and that the
ship could easily be steered by varying the speeds of the rotors,
or put into reverse by simply reversing their direction of rotation. Also the amount of fuel required for covering a given
distance was extremely small.

The two rotors of the *Buckan* had a projected area of
850 square feet, and were sufficient to replace 8,500 square
feet of sail. The dangers encountered on meeting a sudden
storm were thus considerably reduced, and the labour required
to shorten sail in an orthodox sailing vessel entirely eliminated.
In spite of appearances there was also much less danger of
capsizing, for not only were the rotors considerably shorter
than the masts of a sailing ship, but they weighed only 40 tons
as compared with the 180 tons of masts, spars, rigging and sails
required in a sailing ship of the same size.

The success of the *Buckan* led to the equipment of a larger
vessel with three rotors, and it was shown that the new method
of propulsion was suitable for vessels of all sizes. However, the

progress made in mechanically-driven vessels has been so rapid that there has been little opening for ships that rely upon the wind for their motion. It will be noted that the rotor ship depended on a current of air (wind) and so was no better off in a calm than a conventional sailing craft. This ship was invented half a century too late, but attempts were also made to utilize the Flettner rotor in other directions. A form of windmill was designed and built by Anton Flettner, while in America experiments were made in the use of long horizontal rotors in place of aeroplane wings. Neither of these projects seems to have met with the success that was once expected of them.

CHAPTER XIV

The Study of the Winds

THE study of the winds, although it began very early in history
and was of prime importance in navigation, remained for a
long time purely empirical. Less than a hundred years ago it
was still only in an elementary stage. The establishment of a
wind chart by Maury, an American, in the middle of last
century, is a relatively modern achievement and undoubtedly
rendered an enormous service to navigators. The observations
that had been made up to that time proved that there were
zones of regular winds alternating with zones of calm or of
variable winds. Using this information, Maury made a chart
showing the immense stretches of the ocean subject to winds
of a constant direction, like the trade winds, or to regularly
recurring atmospheric currents, like the monsoons of the
Indian Ocean and the China Seas, and indicating the most
favourable and quickest routes for sailing ships. Maury's work
had great repercussions in the commercial world. Before 1850
a voyage of 120 days from Europe to Australia had been considered very rapid, but the data supplied by Maury enabled it
to be accomplished in about seventy-two days. The time taken
on voyages on all the great sea routes was similarly reduced.
Brault's charts followed Maury's and were more complete,
indicating the force of the winds as well as their direction.
Today, the monthly pilot charts published by the Hydrographic
Service of the United States give detailed information on the
probable winds and all the meteorological phenomena of
interest to navigators.

G. La Roërie (Ref. 45), discussing in a very interesting

article the conditions which led to the great maritime exploration of the fifteenth century, examines various hypotheses. In less than 200 years the vast uncharted territories on the maps of the preceding era were filled in and it could be said that the earth was practically all known. He quotes particularly the opinion of the Portuguese Admiral Coutinho, who was very well versed in the history of the great discoveries in which so many of his countrymen distinguished themselves. The admiral considered that it was the knowledge of the winds and currents, of primary importance in the days of sailing ships, which played a decisive part in extending the voyages of discovery. A better knowledge of winds and climates enabled the routes to be progressively lengthened. It seems probable that the Portuguese sailors made a study, half empirical, half theoretical, of the tropical seas and realized that to return to Europe it was necessary to take to the open sea. Columbus, wishing to sail westwards to India, does not appear to have set out on this route by chance; he probably tried to find the zone of favourable winds, on which he would doubtless have received information from some Portuguese navigator.

In any case, the *roteiros* of the sixteenth century (instructions for the use of navigators) show that Portuguese pilots subsequently accorded a very important place to the knowledge of winds. It has sometimes been thought that the Portuguese atlases of the period furnish a further and picturesque proof of this constant interest. The oceans on these maps are decorated with ships under sail, all differently oriented, and some people have wondered whether this was not done intentionally. "The arrangement adopted," says G. La Roërie, "is very unlike that seen, for example, in the work of the Dutch map-makers at the beginning of the seventeenth century. It is obvious that the latter were simply concerned with decoration.

"On the Portuguese plates of a Reynel (beginning of the sixteenth century) each boat is drawn separately and without embellishment. The boats are of a uniform type and their arrangement does not follow any decorative pattern. Is it chance that produced this distribution or has the position of each one been determined by calculation? The idea has occurred to several people that the progress of these boats indicated, at each place, the direction of the prevailing wind. It would thus have been possible to have, in an artistic form, a true diagram of the probable winds."

La Roërie adds that this is an attractive hypothesis but only an hypothesis. Obviously at some points these maps accord with the wind charts of Bouguer at the end of the eighteenth century or with the more modern charts of Maury or of Brault, but it is difficult to establish this accordance with certainty. Moreover, every boat oriented to correspond to the pointing arrows of modern meteorological charts occupies a rather large expanse of sea and its position is uncertain. It is also difficult to ascertain the direction of the wind from the appearance of the boat. This indication is given by the orientation of the sail. "Now when we come to detail," La Roërie continues, "appearances are deceptive because the sixteenth century sail worked in a very different way from that of modern three-masted ships. Often we cannot tell, within forty-five degrees, the direction of the motion of a ship drawn by a scrupulously careful artist at the beginning of the sixteenth century." He considers that there is not sufficient evidence to prove that the old Portuguese atlases give a chart of the winds. Pilots naturally acquired experience which enabled them to extend the length of their voyages, but this came gradually and could not have prompted the first great expeditions.

In his opinion, it was the development of a new form of sail that really made the great voyages of discovery possible. Before the fifteenth century the ships of northern Europe were not suitable for long voyages in the Atlantic. As soon as their dimensions were increased, they became heavy and difficult to handle and they were built almost exclusively to sail before the wind with their sails square. In the Mediterranean, the use of lateen sails enabled the ships to tack, but those of the *Levant* were not suitable for long uninterrupted voyages nor for sailing easily before the wind, so that the Mediterranean sailors hesitated to pass through the Strait of Gibraltar. Voyaging along the coast of Africa, one has to reckon with the regular current of the trade winds and must be in a position to sail along with this current as well as to sail against it. To reach new countries is not everything; one must be able to return from them. The voyages of discovery demanded ships that could sail indifferently with or against the wind.

At the time of the first expeditions along the west coast of Africa in the fifteenth century the Portuguese caravels had at their disposal two interchangeable systems of sails. Lateen sails could be replaced with square sails as the conditions required. That was a notable advance.

The Trade Winds

The persistence of the north-east wind which Christopher Columbus encountered on his first Atlantic voyage made his companions think uneasily about the difficulties of their return in a wind which drove them to the west with such constancy. Later, the navigators of the fifteenth and sixteenth centuries, crossing the different oceans in succession, observed the regularity of the wind régime in the tropical zone, except in the region between south Asia and the equator where the monsoons prevailed. They recognized the existence on each side of the Line of very constant aerial currents of an almost uniform speed, generally between 20 and 25 m.p.h., blowing from the north-east in the northern hemisphere and from the south-east in the southern hemisphere.

English sailors called these constant winds, which were so useful to maritime commerce, *trade winds*. (There is some doubt about the origin of the French word *alizé*, which may be derived from the Spanish *alisios* or from the old French word *alis* signifying regular.)

The trade winds played a large part in navigation in the past. With their help Magellan made the first voyage round the world. It was the north-east trade wind which led the Spanish galleons from the coast of Africa to the West Indies and to Central and South America. From the sixteenth to the eighteenth century, the northern belt of the trade winds was also the route from America to Asia through the Pacific. From the time of Urdañeta in 1565, the trade wind mathematically conducted Spanish sailors from Acapulco in Mexico to Manila in two months.

Although favourable for a crossing, the trade winds by their constancy inevitably made a voyage in the opposite direction interminable and might therefore have been an obstacle to maritime communications. But the old navigators, who knew how to take advantage of them when their direction was propitious, avoided this disadvantage by taking a very different route for the return journey. They then followed the coasts of America, reached Havana and set course for the north until, in the region of the thirtieth parallel, they encountered the west winds which took them to the Azores and finally to Spain. In the Pacific, those who were returning from the Philippines or from China sailed in the direction of Mexico or Peru and, diverting their course northwards, found a longer but more rapid route in the zone of the great west winds. From the north of Japan they sailed in the same latitude until they approached the coasts of California, which they then followed towards the south.

The constancy of the trade winds was of great service to the hardy Polynesian navigators in the course of their voyages round the Pacific. To take their bearings to return to the islands that they had left, they had only to remember the angle at which the wind had struck their sails in the course of their outward journey!

Use of the Monsoons

The credit for having discovered the regular alternation of the monsoons in the Indian Ocean and the advantage that could be taken of them in navigation is generally attributed to Hippalos, a pilot of Alexandria. It is probable, however, that from the very earliest times, the commercial intercourse established by the sea routes between certain peoples living on the shores of that ocean were governed by the regularity of the seasonal winds. There were communications particularly between the Malabar Coast and East Africa and between southern Arabia and Madagascar. "The attraction of the opposite shores," observes Vidal de La Blache, "was all the stronger because there was no anxiety about the return; it was guaranteed by the alternation of the monsoons."

The regular succession of the winter north-east monsoon and the summer south-west monsoon enabled commercial relations to be established between the coasts of China and of

Annam and between the Sea of Jolo and the Celebes Sea.

The Phoenician and Greek navigators were, however, always hampered in their voyages across the Indian Ocean by the monsoons whose régime was unknown to them. H. Froidevaux (Ref. 31) says that, from the time when Nearchus, an admiral of Alexander the Great, followed the coasts of the Arabian Sea and the Persian Gulf between the mouths of the Indus and the Shat al Arab until the reign of Tiberius, navigators sailed from Arabia Felix to India solely by coasting. They hugged the Arab coast as far as Ras al Hadd; they then followed the desolate coasts of Caramanie and Gedrosia to the delta of the Indus and returned by the same route.

E.-H. Warmington (Ref. 73), studying the commerce between the Roman Empire and India, is of the opinion that the tradition attributing to the freedman Hippalos the discovery of the periodical recurrence of the monsoons and their systematic utilization is an over-simplification of the truth. After examining the passage in which Pliny describes the discovery of the monsoons (Hist. Nat. VI, 100, 101–6) and comparing it with other ancient texts, he is inclined to think that even when this phenomenon was known, many difficulties had to be overcome before sailors were able to use these seasonal winds systematically.

The first stage was passed during the reign of Tiberius and of Caligula, when coasting was succeeded by navigation in the open sea. Sailors and traders from the Mediterranean ports, setting out from Aden, which was the rendezvous of the Greeks, Arabs and Indians, or from Hadhramaut, and leaving the coasts of Arabia, took a straight course to the mouths of the Indus and finally reached the old city of Pattala (Hyderabad). There they exchanged their goods for the products of India and China. This new route, although it was shorter and more direct than the old route hugging the northern coasts of the Arabian Sea, was more dangerous. The knowledge of the seasonal winds of India that the navigators of that time possessed enabled them to take the risk but they still did not know all the benefits that could be derived from these winds. A little later, between A.D. 41 and 50, another important stage was passed. Wishing to go farther than the Indus, some hardy navigators took a new and more southerly but safer and quicker route which, from Ras Fartak (the ancient Cape Sygaros), led them across the solitudes of the Arabian Sea to the south of the Kathiawar Peninsula.

Towards the year A.D. 50, the news spread that it was possible to travel from the Bab-el-Mandeb Strait to the markets on the Malabar Coast in forty days. To do this the voyage had to be started at a fixed time in July. Then in the open sea the ship described an arc of a circle to the north; the prow was constantly held into the wind and the rudder and yard skilfully handled. The return journey was made by describing an arc to the south, so that the ship could, if necessary, put in at Socotra. Warmington thinks that this really constituted the discovery of Hippalos. Froidevaux, in the excellent account that he has given of Warmington's work, says in this connexion, "By studying carefully the position of the ports and the state of the sea as well as the geographical data which had been acquired on India and by relating the observations made on the regularity of the winds to the rules of navigation, he came to realize that India extended far to the south and to formulate the theoretical possibility of using the monsoons to sail to the west coasts of the peninsula. Such a discovery does not at all imply that Hippalos himself made the voyage.

"Warmington's theory, based on the texts, shows how the sailors of the ancient world succeeded in turning to practical account a fact of physical geography which they had discovered and of making themselves masters of a secret that the Arabs and Indians had probably known for centuries."

The utilization of the monsoons gave tremendous impetus to trade between the East and the West and soon also with the Far East, for from the fourth century A.D. navigators were able to go as far as China. The Arabs took a larger and larger place in this maritime commerce and their ships ploughed the Indian Ocean, but they were prevented from venturing farther south than Cape Corrientes in the Mozambique Channel by fear of the currents. An Arab, Sidi Ali, in a work on navigation in the Indian Ocean, entitled *Mohit*, which was published in 1554, gave the season of the beginning of the monsoon for fifty different places.

Europeans in the twelfth century had heard that long journeys were made on the Indian Ocean, but as they did not know of the compass probably in use in the East, nor of the regular alternation of the monsoons, they thought that eastern sailors were in the habit, at the beginning of their voyages, of releasing flocks of birds whose flight showed them the route to follow. The travels of Marco Polo and the voyage, in 1486, of Pero de Covilham from Aden to Calicut and thence to Sofola on the coast of Mozambique proved that there was no reason to fear the persistence of the trade winds in the Indian Ocean. Vasco da Gama therefore knew that if he had a propitious wind to take him from Africa to the Malabar Coast, he could be equally certain of being able to return from India to the entrance of the Red Sea.

If voyages were governed by the seasons and undertaken at the same time each year, navigation was easy across the Indian Ocean because the routes had already been mapped out. But when mariners became bold enough to take an oblique direction against the monsoon or to go in face of it, new routes had to be found.

Etesian Winds

Navigators in ancient times were not unaware of the constancy of the wind régime in the eastern basin of the Mediterranean where true monsoons occur. Every morning in the summer the north wind rises on the coasts of Thrace and, blowing furiously under a cloudless sky, sweeps along the Ægean Sea. It drops every evening at sunset and the sea becomes perfectly calm until a very light breeze rises from the south, called *mbatis* by the Greek sailors who take advantage of it to spread their sails.

These regular winds helped to form the psychological unity of Greece and in early times put the Hellenic world in communication with Egypt, making a community of peoples in the eastern basin of the Mediterranean which Homer mentions in his day. Later, St. Paul made his missionary voyages with the help of the etesian winds.

A similar phenomenon also occurs in the summer in the western Mediterranean, where sailing ships often had difficulty in moving away from the Algerian and Moroccan coasts. The crossing from Europe to Africa was always much quicker than the return; the journey from Alexandria to Marseilles took seven or eight times longer than the journey in the opposite direction. This persistence of the north winds in the Mediterranean made Marshall Marmont say that Egypt seemed made to be conquered. Caesar and Napoleon travelled there under full sail, one from the eastern basin, the other from the western. The pirates of Tunis and Algiers took advantage of the north winds to speed them on their return to North Africa.

The Great West Winds

The wind régime which is so regular in the tropical zones is much more variable in Europe, although the predominance of the winds between the south-west and the north-west was observed many centuries ago. Nevertheless, the succession of the winds, blowing from all points of the horizon, seemed for a long time to escape any law and to be purely the product of chance.

In the temperate zone of the southern hemisphere, the constancy of the wind régime of the west is much more marked. This is due to the large expanse of the oceans so that in the little islands which emerge in their midst there is an average of 250 days a year when the winds blow from the western sector. English sailors have given the name of the "Roaring Forties" to the regions of the fortieth parallel in the Southern Ocean, where the *westerlies* or *brave west winds* prevail. Dutch navigators realized that they could take advantage of them to sail to Java by the Cape of Good Hope; the Dutch route passed between forty-three degrees and forty-eight degrees south. In the nineteenth century, particularly, the great sailing ships traversed the zone of the west winds in the southern hemisphere. Since their disappearance the sea routes of the south have again become deserted.

The Danger of Calms

A calm is often worse than a storm for a sailing ship, not only because it delays the voyage and fouls the hulls but because it has its own peculiar risks. Ships becalmed, as sailors call it, and no longer answering to the helm, are the prey of the sometimes enormous swell; everything in the rigging may be torn away and they are often in danger of capsizing. The zone of equatorial calms, lying between zero degrees and twelve degrees north and varying slightly within those limits according to the season of the year, was very formidable in the days of sailing. This region at the boundary of the two hemispheres, where the absence of wind sometimes kept ships stationary for days in succession was called the "black pot". The rain falls in cataracts from a sky darkened by large black clouds; the air is moist and the heat overpowering. The crew had to exert every effort to take advantage of the lightest breeze, but very often there were days of flat calm without a ripple on the surface of the sea, when the ship swayed at the mercy of the swell, her sails flapping pitifully along the masts. Occasionally a ship would remain becalmed as long as a month, but this was exceptional; generally the "black pot" was passed at the end of a few days. Sometimes an absolute calm at the equator would be suddenly broken by violent squalls capable of dismasting the stoutest vessel if the crew were taken by surprise.

The width of the zone of calms is not the same all along the equator. In the Atlantic they are particularly persistent and sailors generally tried to cross the Line towards thirty degrees longitude where they are the least extensive. The wind charts gave them useful information on all the points of the ocean where the calms were most to be feared. On the coasts of Chile, for example, ships, after loading up with nitrate, might be detained for several days before they could set sail on their homeward voyage. There are also zones of calm in both hemispheres between the domain of the trade winds and the domain of the great west winds and I remember on one occasion returning from Dakar and seeing, when the steamer stopped to revictual, a large Portuguese sailing ship immobilized half way between Madeira and Casablanca. The exhausted crew had come to the end of their food supply after waiting two weeks for a favourable wind to take them to the Moroccan coast.

The enormous swell near Cape Horn made the calm particularly dangerous there. Although it generally lasts only a few hours, it could cause great damage to stationary ships. Lacroix mentions unusually long spells lasting three to four days; three-masted ships, like the *Ellenbrock* in 1885, caught between two enormous waves, have been known to capsize in a flat calm.

The routes were dotted with many islands which served as ports of call. At some, like the Cape Verde Islands, St. Helena and Ascension Island, ships found food and water while they waited for favourable winds.

Difficulty of Access to Certain Coasts

In seas subject to a régime of prevailing winds, access to a large number of islands from the side struck by the wind is practically impossible, as it is, for example, on the west side of most of the sub-Antarctic islands. In the domain of the trade winds it is generally the east side which is inaccessible unless, of course, there is some indentation on the coast which forms a natural harbour. The ports of many tropical islands are therefore found on the leeward side. In the Antilles, Fort-de-France on Martinique and Port of Spain on Trinidad are on the west. In the New Hebrides, where the trade wind is particularly violent from May to September, the coasting ships are often unable to put in on the east side of the islands for several weeks in succession. The same disadvantage is found in the zone of the monsoons, where certain islands are periodically inaccessible for several months on the side where the wind blows.

There are, of course, islands off the coast of Europe which may be temporarily isolated during strong gales—not to mention lighthouse-keepers sometimes cut off from the mainland for weeks by the raging sea!

Where there are no sheltered bays along an exposed coast, it is important, in choosing a site for the construction of a port, to consider the direction of the prevailing winds. Even in a well-planned port, shipping may be hindered or interrupted by the violence of the wind, as it is at Marseilles on days when the mistral is blowing strongly.

The wind also piles up ice-floes at the entrance of some ports, making them inaccessible in the winter. Sydney in Nova Scotia, where normally the sea does not freeze, is regularly blocked for several months, until late in the spring, by ice brought by the north-east winds from the St. Lawrence and from the bays of Newfoundland. St. Pierre and St. Johns are not so frequently affected.

CHAPTER XV

The Wind and Aviation

THE study of the atmospheric circulation has become very important since the development of aviation and the establishment of trans-continental and trans-oceanic commercial air lines. Aircraft have an even greater need than shipping for an accurate weather forecast. The winds, in fact, directly affect air transport; like the great west winds over the North Atlantic between Europe and America they can facilitate or hamper it.

Aircraft

The true speed of an aeroplane is its speed in relation to the ground in an absolutely calm atmosphere. It is the speed which it possesses in relation to the mass of the ambient air, whether that air is animated or not by a movement of progression. It is usually different from its absolute speed which is the real speed of the aeroplane, whose progress is slowed down if it has the wind in face of it or increased if it is moving before the wind. Moreover, the wind is seldom parallel to the route followed; usually the aeroplane receives it obliquely. The pilot has to take into account the drift caused by a lateral wind, which is, besides, rarely horizontal, but generally ascending or descending.

The development of aviation therefore requires a large number of wind observations, carried out by means of kites and more often today by means of pilot balloons and radio-sondes. They enable the wind régime to be determined at an altitude of generally between 13,000 and 23,000 feet. The aerial currents in the upper atmosphere are not necessarily the same as those

L   161

near the ground—often they are very different and there are superposed layers of wind having different directions. Again, at a given altitude, the direction of the wind is subject to periodical variations and changes which depend on the weather. It is of great assistance to the pilot to know all these changes, as well as the average wind speed at different altitudes. The latter varies according to the wind direction and the season. Usually it tends to increase with height.

Even in very good weather when the atmosphere is apparently calm, it may be subject to violent eddies and air pockets which give no warning sign. These eddies make the machine vibrate when it encounters them and a few years ago caused three aeroplanes in succession to crash in Colombia as they were preparing to cross the Andes.

The speed and direction of the winds in the stratosphere were until recently quite unknown. The stratosphere begins at an altitude which varies in different places and at different seasons but which is on an average about six or seven miles above the earth's surface. It was generally thought that the air at those altitudes was calm, but recent investigations have revealed the existence of very strong currents, reaching 156 m.p.h. The study of the winds of the stratosphere has become very important in the last few years in preparation for the time when ever faster aeroplanes at greater altitudes will make stratospheric flights.

At medium altitudes it is important that the pilot should be kept informed not only of the general probability of the weather but also of the details of passing turbulences, especially of squalls. In some cases observations of the atmospheric situation must be communicated to him hour by hour. A number of meteorological stations have therefore been set up along the routes of the great airways. The establishment of airports also necessitates a very exact knowledge of local wind régimes. We may mention, in this connexion, the danger of certain ascending currents which can seriously hinder the landing of dirigibles. Some Zeppelins have been obliged to wait several hours before they could land, though these facts are now of little more than academic interest.

The airways of the North Atlantic, more than any others

in operation, whether the route passes over Newfoundland or the Azores, require the most minute data on the atmospheric conditions and the movement from west to east of the depressions which are so common in those latitudes. The crews on transatlantic flights are advised at their departure of the approximate position of these disturbances. They can be informed by radio in the course of the journey of the progress and development of depressions and can consequently change their course northwards or southwards, according to circumstances, to avoid adverse winds as much as possible. On the lines from Europe to the Far East, aeroplanes have to be warned, according to the season, of the intensity of the monsoon in India and of the formation of typhoons.

Accurate weather forecasting enables pilots to avoid a snowstorm while flying over a mountainous district, a stormy gale over the equatorial forest and a sandstorm in crossing the desert.

Flying over desert countries, particularly the Sahara, demands special precautions because the dust-laden winds considerably reduce visibility. It can become nil in the horizontal sense, without necessarily being so vertically or obliquely. Sand storms are not always so dense as to prevent pilots from distinguishing objects on the ground through the less thick parts of a cloud of dust. Landing in a sand storm, however, is always dangerous, especially in the neighbourhood of a "wall" of sand; the aeroplane may be forced to pancake by the descending currents behind whirlwinds with a horizontal axis. In the Niger Colony, between Bilma and Agades, flights are made at night to avoid the ascending winds laden with sand during the day. Nevertheless, modern aeroplanes which can rise rapidly to 13,000 and 16,000 feet are able to fly over most sand storms in a clear atmosphere.

Gliding

Gliders—flying machines with no engine—brought the wind into a novel and interesting use by the practice of gliding, which consists of taking advantage of ascending currents to gain height. The inventor of this sport was a German, Otto Lilienthal, who made in this way about two thousand flights between 1881 and 1896, several years before the Wright brothers made their first flight in an aeroplane—in December 1903.

At the beginning, glider pilots succeeded in keeping in the air only along slopes and were to some extent confined to the same sort of terrain. Gradually they learnt how to climb below clouds (cloudy ascents), in the face of storms and even in stormy clouds. Gliders, launched by means of a rubber extensible spring or a winch, took advantage of "slope ascents" to rise uphill; they were thus able to reach the clouds and to use the favourable ascending currents that they provoke.

The Germans, who are masters of the art of gliding, have carried out extensive research on it. In 1929, an Austrian, R. Kronfeld, made a flight of 100 miles. German pilots shortly afterwards surpassed this distance.

The discovery of a new means of ascent, called a "thermal", which takes advantage of the fact that the air, heated by the sun at certain points, rises in narrow columns and produces rather strong currents, enabled gliding to make great progress. The pilot who finds these columns rises inside them, and then tries to find another column without losing too much height.

R. Kronfeld (Ref. 42), in his interesting work on gliding, stresses the importance of the wind. The glider pilot depends entirely on the winds and on his ability to make use of them. Obstacles on the earth's surface divert the wind from its regular course and the slightest irregularities cause the formation of eddies which may reach heights of 7,000 and 10,000 feet. Gliders have been steadily improved and as they are now able to take advantage of very slight ascending currents, they can make longer and longer flights.

Many countries realized that gliding was an excellent means of giving young men a taste for, and a sense of, the air and that it could form a valuable part of the training of aeroplane pilots. In Germany and the U.S.S.R. particularly, great efforts were made to develop it.

The practical value of gliders has been rather disputed. Gliding enthusiasts claim that it is a cheap and safe means of encouraging a large number of young people to take an interest in aviation. Others consider it nothing more than a sport. In point of fact, gliders are very useful in making aerological measurements and in studying atmospheric movements. They enabled experiments to be made over Berlin to determine the aerial currents in the vicinity of the aerodrome at Templehof. At the beginning of the flight, the gliders were towed by an aeroplane—a method which is being increasingly used —and were released when they had gained the desired height.

Argument about the practical value of gliders, which in their more navigable form are known as sail-planes, was finally settled during the Second World War when motorless aircraft were used extensively for the transporting of troops. These aircraft were towed by a powered aeroplane until near the intended point of troop landing when the tow ropes or cables were cast off and the motorless aircraft landed, the powered aircraft flying away. Arnheim and Normandy were two places where this practice earned fame.

An Example of Passive Obedience of the Wind: Balloons

The first ascent in a balloon made by Pilatre de Rozier on 21st October, 1783, roused great enthusiasm. The problem of rising in the air had been solved. The means of directing balloons still had to be found. One of the difficulties was that their large surface put them at the mercy of the slightest breeze. After 1783, a number of aeronauts and engineers experimented with various devices to steer balloons. For a long time their efforts met with no success—the dirigible balloon is a relatively recent achievement dating only from the beginning of this century. All the methods proposed proved useless as soon as the wind speed exceeded 4 to 7 m.p.h. Towards 1860, it was realized, after many failures, that the science of mechanics was not then sufficiently advanced to offer any solution to the problem.

The art of piloting spherical balloons therefore consisted of choosing the direction of their course to take advantage of favourable currents which would bear them along. Commenting on the unsuccessful attempts to free balloons from their servitude to the wind, Victor Hugo, in a letter written to Nadar in 1865, whimsically remarked, "The balloon is a perfect example of passive obedience to the wind. The balloon is judged and condemned. Let us, however, make one important reservation. Where the direction is already indicated, the balloon can be useful. If the wind takes charge of the route, if its breath is the pilot, the balloon, with its specific lightness, is a suitable means of locomotion. There are some fixed winds. Two bands of winds, the polar band and the equatorial band, turn incessantly round the globe. These two bands meet and cut each other at right angles at the equator. Hence, towards the tropics, those sharp rents in the atmosphere, those sudden fissures in the region of calms from which cyclones and typhoons rush headlong. These two circles of wind wait for the aerial navigator and mean for him a task already accomplished."

Pilots using the atmospheric currents at the right time have made very good journeys in balloons, but these are generally purely sporting feats without any practical application. It is interesting, however, to recall the tragic attempt made by the explorer Andrée in 1897 to reach the North Pole in a balloon. It was certainly a daring project to try to follow a fixed course, but Andrée counted on being able to steer the balloon, at least to some extent, by means of a guide-rope fixed to the nacelle and of sufficient length to trail on the surface of the sea or the ice-pack, thus providing a point of support to the apparatus. He thought that the friction of this cable would reduce the speed of the balloon to a speed lower than that of the wind. A sail could then be used, which would enable the balloon, if necessary, to follow a course at an angle of about twenty-five degrees with the direction of the wind. He also thought that in the summer the aerial currents in the Arctic basin blew from the south.

On 11th July, Andrée and his two companions left Port Vigo at the north-west point of Spitsbergen in the *Oernen*, expecting to take about thirty hours to cover the 705 miles separating them from the Pole. The wind was then south-west, but the favourable meteorological conditions which Andrée had anticipated were not maintained and the novel device for directing his balloon in which he had placed so much confidence became detached and fell into the sea shortly after he set out. The *Oernen* reached the eighty-second parallel without difficulty. Two days later he sent a message by carrier pigeon to Spitsbergen that he was still at this latitude but had drifted fifteen degrees east. After that, there was no more news. But on 6th August, 1930, the *Bratvaag*, a small Norwegian ship on a seal-hunting expedition, finding the position of the ice exceptionally favourable, passed the strait between Spitsbergen and Franz Josef Land. At White Island the crew found the remains of Andrée's expedition and his journal. It was thus learnt that he had had to abandon his balloon on 14th August, 1897, after landing at 84° 32′ N. and 30° E.

The Kite

The idea of being lifted up and carried by the wind is very old and is based on the direct observation of nature. The withered leaf flies with the wind and is carried far. As soon as the wind stops, the leaf falls to the ground. Imitating nature, man launched multi-coloured squares and variegated polygons into space. When he thought of maintaining them by means of a cord so that they could withstand the wind, the kite was invented. The pressure exerted against them by the wind made them rise, and the first flying machine conceived by man floated in the sky.

The kite, which commonly consists of a piece of paper or cloth stretched on a light wooden quadrilateral frame, is a well-known children's toy, but it is significant as representing a very ancient utilization of the wind. It is not only a simple plaything but has had for a long time various practical applications.

In the north of Melanesia, particularly in the Solomon Islands, the natives use kites made of banana leaves for fishing. Attached to the canoes with fine tropical creeper and kept about sixty-five feet above the water, they support a line with a hook on the end of it which runs on the surface of the waves. A small ball, made of spider's web, sometimes takes the place of the hook at the end of the line and serves at the same time as bait.

The kite was used by Benjamin Franklin in 1750 for his memorable experiment on atmospheric electricity. It was used for the first time in 1749 in meteorological research, but all that it was then required to do was to raise minimum thermo-meters into the air. Systematic wind observations with the aid of the kite have been made only since 1894. In that year, a barothermograph was sent up about 1,640 feet. Great progress has been made since then and modern meteorological kites, which can reach an altitude of over four miles, have the great advantage of maintaining recording apparatus at a practically constant height. For investigations in the upper atmosphere they are generally used in tandem. The principal kite supports the instruments and is maintained by a steel cable. When the weight of the cable slows down its ascent, the second kite is sent up, but heavier cable is used because this second apparatus adds its pull to that of the first. Other kites are then arranged at points along the steel cable, their number varying according to the force of the wind.

It was formerly the custom in eastern Asia to use kites representing dragons as signals and ensigns of war. The same custom existed in Roman times and lasted until the beginning of the Middle Ages. P. Thoene (Ref. 70) records some curious anecdotes showing that in very ancient times the kite served military ends. In Japan, in the third century B.C. when the government was in the hands of a despot, some rebels made a plot to enter the Imperial Palace. It was on this occasion, if the legend can be believed, that one named Han Sin invented the kite. Taking advantage of a favourable wind, he launched his device, in the form of a winged ogre, into the air and made it hover over the Imperial Palace. Then, by means of the cable which held it, he measured the length of the tunnel that would have to be made to reach the emperor's apartment. Japanese chronicles record that men mounted on enormous kites flew over an enemy camp to reconnoitre the positions of their opponents, as military aeroplanes do today. In the reign of the third Tokugava-Shogun—about 300 years ago—the scholar Yuino Shosetsu appears to have mounted on a kite to fly to the Imperial Palace of Yedo (Tokio). From the air, he estimated the chances of combat, then returned to the rebels' camp. The sight of the man carried by a kite frightened the emperor and all his court to such an extent that he made a law prohibiting, on pain of death, the construction of a kite exceeding the dimension of four leaves. Later, the "aviator" Yuino Shosetsu was arrested and forced to commit suicide. Finally, there is another legendary feat, also recorded by P. Thoene, which is said to have been accomplished by the rebel chief, Ishikava Goyemon, living in Japan at the end of the sixteenth century. To finance his revolutionary schemes he wished to steal the two golden dolphins which decorated the tower of the palace of Nagoya. Mounted on a large kite, he reached the dolphins and stole their golden fins. Later he was caught and boiled in oil with all his family.

CHAPTER XVI

Benefits and Drawbacks of the Wind

THE two principal uses that man has made of the wind are as a propulsive force for sailing ships and as a driving force for windmills, but these are not its only uses. It still serves for other, and often very unexpected, purposes. In some circumstances it not only cannot be used but is a dangerous force against which man must defend himself.

The Use of the Wind in Industry

Except in a few limited cases, the wind is not used in industry today to provide motive power although it is sometimes used in other ways. One of the oldest examples is provided by the Romans, who invented the "wind bag" for their metal-working. It consisted of bellows made of skin and fitted with a diaphragm and was operated by a man. Th. Rickard (Ref. 66) quotes the English author, John Farey, writing in 1811 on the subject of lead-mining in Derbyshire in the time of the Romans: "The mineral was treated at the top of high hills on wood fires stirred up solely by the wind. The miners surrounded the fire with heaps of stones; they probably arranged openings so that the wind had a more direct action on the fire." Rickard thinks that the dependence of metal-working in Roman times on the most frequent wind is further corroborated by the slag heaps showing the site of the ancient iron-smelting furnaces along Hadrian's Wall, which are all situated on high hills exposed to the west winds.

The Indians of Peru at the time of the Spanish conquest employed a similar process for the extraction of silver. Ignorant of the use of bellows, they placed their very primitive little clay furnaces (*guayras*), similar to flower-pots pierced with holes, on the side of the mountain facing the prevailing wind. Rickard says that Acosta claimed to have seen 6,000 *guayras* on the slopes of Potosi in 1608 and that De Léon wrote in 1547 that there were so many of them that the mountain seemed to be illuminated. To increase the current of air, especially for refining, the Peruvians blew into copper tubes.

The metal workers in the Ardennes formerly used the Catalan furnace, provided at ground level with a narrow pipe ending in a funnel into which the wind rushed to stir up the fire. Before these wind furnaces could be set up, sites had to be found on the hillside exposed to the prevailing south-west wind.

Today air currents produced by powerful blowers are used in the fusion of metals, the wind as it occurs in nature being too capricious for the needs of industry.

The ease with which moving air causes evaporation makes it eminently suitable for drying a large number of products. Either natural or artificial ventilation is used for this purpose. The wind plays a particularly important part in the preparation of the salt in salt-pans. The most productive are situated in very windy coastal districts with a low rainfall, like those in the Mediterranean Basin (south of France, Sfax, Port Said) and in the domain of the trade winds (Cape Verde Islands, Curaçao, Aden, Djibouti, Diego-Suarez, etc.). The French salt-pans in the south from Var to Pyrénées Orientales are more favourably placed than those on the Atlantic coast, not only because the temperature there is higher and the rainfall lower, but because they are subject to the mistral—an extremely drying wind.

Continuous ventilation has to be provided in many mines during working hours to cool the air when it is too hot and to renew it when it is tainted with dangerous gases. In the absence of a strong natural current of air, artificial ventilation is produced by powerful fans. In some gaseous mines, air currents charged with an inert body, generally pulverized shale, are circulated in the galleries to reduce the risks of explosion.

A rational system of ventilation in workshops and factories is indispensable to cool and purify the air, as it is in the engine-rooms and stokeholds of ships, which are ventilated by means of air-shafts placed on deck and oriented according to the direction of the wind.

In some towns, Lille in particular, an ingenious method of ensuring the natural ventilation of subterranean passages has been devised by means of "wind traps". This method, advocated by J. Chappuis (Ref. 19), uses the natural energy of the wind and the traps employed to capture it consist essentially of a grating placed over each of the vent-holes. The bars of the grating, instead of being of square section, are like those used in Venetian shutters. They are parallelepipeds with angles of forty-five degrees. In the latest models the bars have been replaced by parallel metal strips inclined at an angle of forty-five degrees; the first series of strips is supported by a second, formed of strips parallel to one another and perpendicular to those of the first. The gratings are placed on the pavements at the opening of the two shafts ensuring the ventilation of the passage. With this device, it is possible to capture part of the energy of the wind along the ground (about 35 per cent) either to drive the ambient air into the vent-hole or to draw out the air contained in it. To make this system of wind traps fully effective, both ventilating shafts must be cut off level with the roof of the tunnel, since they serve, according to the direction of the wind, either as inlets or outlets for the air and must, in any case, evacuate the gases from the tunnel.

The Wind and Town Planning

The wind purifies the air of large cities and industrial areas. The quantity of carbon-dioxide daily produced by the respiration of the inhabitants of a large city and the still greater quantity proceeding from domestic fires and combustion of all sorts amount to millions of cubic feet. The air would become still more unwholesome than it does if it were not kept continually in motion. When, fortunately very rarely, there is total stagnation lasting for some time a very dangerous condition may arise. This happened in an industrial valley in the neighbourhood of Liège in Belgium about twenty years ago, when a very thick fog caused a number of deaths. It is also the absence of wind which gives rise to the thick fogs in Paris and more particularly in London on certain days in the autumn.*

The wind fortunately has the effect of carrying away the smoke and all the impurities of the atmosphere of a large city. An exact knowledge of the wind régime is therefore indispensable in planning the lay-out or extension of an urban district of any size. Obviously, factories and buildings which will discharge smoke, dust or chemicals into the atmosphere should not be placed on the side from which the prevailing wind comes but on the side from which the wind blows least frequently. The impurities that they discharge will then generally be carried away to the neighbouring countryside and the atmosphere of the town will suffer the minimum of pollution.

The direction and frequency of the wind are not the only considerations. The town-planners must take into account another factor—the turbulence or eddying motion of the air which diffuses the smoke and gases thrown into it. The stronger the turbulence, the more quickly will they be dissipated. This eddying motion depends on the wind speed and on the state of vertical equilibrium of the atmosphere (thermal gradient).

C.-E. Brazier (Ref. 10) studying the climate of Paris, where the prevailing winds are from the western sector, followed by those oscillating between the north and the north-east, shows the advantage of considering the direction of the wind in the orientation of the streets in any urban district. Streets placed parallel to the most frequent wind are well ventilated and the smoke—as well as the gases escaping from motor vehicles—are automatically swept away. In the summer this arrangement will make the heat more bearable. If the streets are perpendicular to the direction of prevailing wind, part of its force will be lost, but the cold in the winter will not be felt so keenly.

It is a well-known fact that the circulation of the air, in the absence of wind, is not the same above large cities as it is in the open country. This is due to the conservation of heat in the buildings, the rapid drainage of the rain-water and the impurities in the atmosphere. A German architect, M. Grunow, studying the effects of these factors on the natural circulation

---

*Translator's Note.—The effects can be very serious in London. A particularly noxious fog occurred in December, 1952, when the number of deaths recorded in one week was 4,703, compared with 1,852 in the corresponding week of the previous year. Most of this difference could be attributed to the fog.

of the air, has shown that the laws to which they conform are of
particular interest in town-planning. In calm weather, the
dust and smoke from a town may rise 650 or 980 feet. The
particles in suspension in the atmosphere absorb up to 30 per
cent of the solar radiation, but, on the other hand, they reflect
the heat radiated from the ground. The heat discharged by
the fires in a large city slightly increases the temperature of the
air. For Vienna, this increase would be, according to M.
Grunow, 1.5° C. These various causes combined are capable
of producing a local rise in temperature of the order of 8° to
13° C., with the formation of an ascending column of air with
an intake of air at the base. In moderate winds the column is
carried away. This architect proposed, for the arrangement of
future urban districts, a scheme specially designed to facilitate the renewal of the air with the streets forming, in plan,
equilateral triangles whose apices would be occupied by
circular open spaces.

Use of the Sail on Land

Wind energy, captured by means of the sail, has been used
very little for transport on land, where this kind of propulsion
would generally meet with too much resistance. Only very
rarely have vehicles been provided with sails to take advantage
of the help of the wind. The Chinese a long time ago had the
idea of erecting them on their wheelbarrows when the direction
of the wind was favourable. In Newfoundland, in the winter,
they are fitted to the sledges used for the transport of
wood.

An amusing attempt to use the wind on railways was once
made in America and reported by the *Revue Britannique* of
January, 1830, in these terms: "An experiment has just been
made at Baltimore in the United States. We know that in
some northern countries sails are fitted to sledges to increase
their speed when the wind is favourable and that these vehicles
sometimes cover great distances with incredible rapidity.
Reasoning by analogy, it has been supposed that sails could
also be used successfully on such smooth and glassy surfaces
as railway lines and that, in this way, expensive fuel could be
saved. A railway coach was therefore fitted with a large sail

and put in the charge of an experienced navigator. The coach, when the sail caught the wind well, was driven at an extraordinary speed of nearly twenty-one miles an hour. Although the sail had been suitably furled and even lowered, the impetus was so great that it was difficult to stop the coach." The experiment may make us smile, but it was not ridiculous if we remember how light the first railway coaches were. However, in spite of its brilliant success, this original utilization of the wind was not developed.

Ingenious sportsmen have taken advantage of the propulsive force of the wind by fitting large sails to very low, light sledges, on which they can move rapidly over the surface of a frozen lake. This sport has had a great vogue in Russia, Canada and the United States. It is also a popular amusement in the winter among the inhabitants of St. Pierre and Miquelon, who use "ice boats" in which they can perform manoeuvres on their frozen ponds.

"Sail cars" are designed on a similar principle, but the runners are replaced by wheels and they are used, not on ice, but on beaches of hard sand, such as are found in Belgium, Brittany and at Berck in the Pas-de-Calais. In Egypt a few years ago, officers of the Royal Air Force formed a "Sand Yacht Club" at Abu Sueir, a little to the west of Ismailia on the road from Port Said to Cairo. They used the fuselages of old aeroplanes fitted with wheels and sails to move over the desert sand.

Domestic Uses of the Wind

In the country, where there is often reason to complain of the wind flattening crops, shaking orchards and scattering haystacks, it does, on occasion, render small services in helping to winnow the grain and move the scarecrows to frighten the birds away from the gardens.

In some districts, special virtues are attributed to the currents of fresh air passing through certain natural caves. In the Pouilles, they are considered to improve the oil. The vine-growers of Mendrisio and Capolago, near Lugano, place their wine in the screes of Monte Generoso because they think that the fresh air circulating through these enormous heaps of rock improves its quality. The vine-growers of Orleans used to put weak or insipid wines in fresh and airy caves where they acquired a "piquant and appetizing flavour". The cheeses are left to ferment in the caves of Roquefort. These caves, called *fleurines* and situated among crumbling rocks, are traversed by a continuous current of cold damp air which promotes the fermentation of *penicillium* and gives the famous cheese its creamy consistency.

A certain number of caves in the Juras are traversed by very cold currents of air which, in the middle of the summer, can freeze the water oozing along the chalky walls. In one at La Givrine near the St. Cergue Pass, I saw ice formed in this way when outside the temperature was about 20° C. There are four of these natural refrigerators in the Department of Doubs —at Pierre-Fontaine, Luisans, Arc-sous-Cicon and Grâce-Dieu. The last, near Chaux-les-Passavant, was famous in the seventeenth century and its ice was commercially exploited.

Every country offers some examples of the domestic uses of the wind. In Norway and in Newfoundland, it dries the fish, which is hung up, placed on wicker trays or spread out on the beach for this purpose. The inhabitants of the Magdalen Islands in the Gulf of St. Lawrence sometimes commit their letters to it in a barrel fitted with a sail to take it to the mainland. A postal service entrusted to the wind, whose caprices are so well known, is, to say the least, hazardous! They also cool their drinking-water by hanging up water-bags in a current of air. We must not, of course, forget the special use that is made of the wind in Tibet to drive the prayer mills turning on the roofs of Buddhist monasteries.

When the wind blows through hollow rocks on the sea-shore it may produce strange, unearthly sounds which stir the emotions of those who hear them. The coast of Cornwall has many cavernous rocks and the wind, when it blows from a certain direction and a storm is approaching, makes, according to local legends, "a deep and mournful cadence, like the tones of an organ. No fisherman puts his boat out to sea when these ominous voices are moaning."

Wind instruments obviously have only a distant connexion with human geography, but it is amusing to note the use made in different countries of natural currents of air to obtain more or less musical effects. The Aeolian harp, which has nothing in common with the real harp except its name, at one time had a certain vogue in England. It was usually installed in gardens and had eight to ten strings which became resonant in a current of air and produced harmonious chords.

Some millers in Portugal hang a series of earthenware vessels, in order of size, between the sails of their windmills. The vessels produce a rather pleasing sound, modulated with the variation of the angle of attack of the wind. This custom of furnishing the sails of mills with musical instruments appears to be rather local. It may be noticed particularly on the cliffs of the Cape da Roca at the western end of the Serra de Cintra.

There are also the wind bells of Tibet and the whistling pigeons of China and Turkestan. The latter carry under their tails a whistle worked by the displacement of the air caused in their flight.

Precautions Imposed by the Wind

Engineers planning the construction of buildings have to take into account the considerable and intermittent pressures exerted by the wind. Gusts particularly produce sudden and violent stresses. L. Cattala, in a very well-documented article, shows the necessity for laboratory experiments, which are the only means of calculating them accurately. Experiments of this kind are generally made in a wind tunnel as well as in the natural wind. Between 1884 and 1890, when the Forth Bridge was being built, B. Baker made some measurements with steel plates of various dimensions placed perpendicularly to the direction of the wind. More recently at Zeebrugge in Belgium wind stresses were studied on a steel lattice tower. Two other towers, placed on either side of it, supported the cables. At Liège, experiments were carried out on fixed panels on the face of a building. "It is fallacious," Cattala says, "to try to estimate the forces exerted by the wind on a building on the basis of its effects on a flat plate. The aerodynamic forces on a whole structure are not the sum of the stresses on the various parts studied separately. The stresses of the wind on any part of an obstacle and on the whole of it depend on the shape of the whole

M

obstacle." Moreover, measurements in the free air are difficult and often uncertain owing to the complexity of natural phenomena. The best procedure is to make experiments on models in a wind tunnel.

To the rapid variations of the pressure exerted by the wind, Cattala continues, is added the variation in the speed of the gusts, which is greater still in tropical cyclones than in storms in temperate zones. It has even been possible to record a certain rather regular periodicity. Engineers must, therefore, without knowing the period to avoid, do what is necessary to prevent resonnance in buildings such as suspension bridges, radio towers, etc., when a cyclone occurs. German technicians of the Telefunken Co. applied this principle to the construction of radio towers on the coasts of China in 1925.

In 1903, an electrical device was installed on the railway bridge at Levens, near Ulverston, to measure automatically the speed of the wind by the pressure that it exerted. When the pressure exceeded 32 lb. per square foot, the instrument barred the approach to the bridge, thus preventing the trains from incurring the danger of crossing it in a high wind.

Electrical transmission lines are also affected by the wind and consideration must be given in their design to the atmospheric conditions of the regions that they will cross and particularly to the prevailing winds which will exert a pressure on them. The spacing of the conductors, which may deviate under the influence of the wind about thirty degrees either side of the vertical, has to be calculated. These precautions are particularly necessary in very windy countries if long spans have to be made to cross ravines because air streams act like eddies on the cables submitted to their stresses. It has frequently been observed that conductors make contacts during storms. In France, regulations are in force to control the spacing of high tension lines so that contacts are not made by the action of the wind.

The Spread of Fire

The wind by fanning the flames promotes the spread of fire and is often responsible for the extensive damage caused in so many towns and villages when a fire breaks out. Many disastrous fires could have been avoided if they had occurred in calm weather, for then it is much easier to keep them within limits.

The risks are particularly great in very windy districts where the buildings are mostly made of wood. In Central Rusisa, the thatch-covered cottages generally burn one after the other when a fire breaks out in a village, even if there is only a slight breeze. To avert this danger the thatched roofs are now being replaced more and more by sheet metal, less aesthetic but much safer. On the very windy islands of St. Pierre and Miquelon, where wooden houses are common, the chief town has several times been partially or totally destroyed by fire; the use of wood was therefore forbidden about fifty years ago for buildings in the centre.

The great fire in Chicago in 1871 was a very striking example of the disastrous effects of the wind. The town was then only thirty-four years old and already had 300,000 inhabitants. On the evening of 8th October, a fire broke out in a stable. It was rapidly spread by a wind blowing in violent gusts from the south-west and scattering sparks which started new fires nearly everywhere in the area since most of the houses and large commercial buildings were made of wood. The fire lasted two days until the wind dropped and the rain began to fall. It claimed several hundred victims and caused 200,000,000 dollars worth of damage; 20,000 houses were burnt down and 100,000 persons were homeless as a result of it. One curious fact was that the ashes from this enormous conflagration were carried so rapidly by the wind that they were picked up four days later in the Azores. In the same year, 1871, fires devastated large areas of prairie and forest land in the surrounding districts, principally in the States of Wisconsin, Michigan and Minnesota, causing still greater loss of life. The worst was in Wisconsin, where 1,500 persons were burnt. The fire was spread by a tornado which raged for half an hour in the morning of 8th October; there was no rain at the time and it followed a prolonged period of drought.

If there were no wind, forest fires would be much less serious than they are and there would not be occasion to deplore the destruction of immense stretches of woodland every year.

In France alone, hundreds of acres of forests are destroyed nearly every year in the south and the south-west. Plantations of conifers are frequently affected and the wind is chiefly responsible for spreading the fire to neighbouring forests, rekindling it when it is on the point of extinction and rendering ineffective the struggle to keep it in check. Torrents of sparks and pieces of burning wood fly in the air. The fire generally burns itself out or stops when the wind drops or changes. In some cases, when it is blowing in a suitable direction, counter-fires can be started over a length of a few miles. A sudden change of wind, however, may spread the counter-fire to the area that it is intended to protect and the fire-fighters, helplessly surrounded, perish in the flames.

Forest fires stirred up by a dry and violent wind sometimes spread so rapidly that all the efforts made to check them are absolutely useless. Nothing can arrest their progress in Provence, Maures and Esterel when the mistral is blowing. They are just as uncontrollable in the pine woods of the Landes when the wind comes from the west and in the Alpine forests during the *föhn*.

In Canada, large areas are the prey of the flames almost every year. Forest fires are particularly menacing in the month of May and during the periods of droughts and heat waves which occur every summer in the North American continent. I remember seeing, in the west of Canada in August 1936, enormous stretches of sky veiled by the smoke from the fires then raging in the Rockies.*

Special preventive measures are taken in many countries to avoid them. In the forests in Germany, long, narrow cuttings are made, oriented N.W.-S.E. and proceeding in parallel lines towards the south-west, which is the direction of the prevailing winds. In Australia, particularly in Victoria and New South Wales, the struggle is well organized. The danger is greatest in January and in the first fortnight of February, which is the hottest and driest time of the year. Fire is combated with fire and corridors are made through the heart of the forest to facilitate the use of counter-fires at the right moment.

*Translator's Note.—On 26th September, 1950, the sun appeared blue in Donegal and in other places in Western Ireland through the smoke borne across the Atlantic from the forest fires in Canada.

The tropical forest, generally very damp, is much less menaced by accidental outbreaks than the wooded areas of temperate zones, but there is a risk on the edges from the bush fires started by the natives for various purposes. On Madagascar, they set fire to the savanna every year to promote the growth of new grass and obtain better pasture for the herds of cattle. In West Africa, bush fires help the natives in their hunting. Under the influence of the wind, these fires, started deliberately, sometimes assume disastrous proportions, destroying numbers of animals, cutting off communications and threatening villages and crops. They are not confined to the savanna but every year encroach farther into the neighbouring forest so that in some districts on high plateaux and in the west of Madagascar the forest is in danger of totally disappearing.

The herdsmen of the Landes used to set fire to the heather to improve the quality of the grass, but the flames, driven by the wind, very often destroyed the pine forests at the same time. The Arab shepherds, by thoughtlessly burning the dry grass on their pastures, destroyed vast forests of cork-oak in the mountains of Algeria. Settlers in Canada breaking new ground are in the habit of burning the felled trees but they cannot do it without a special authorization. The "licence to burn" is given to them by an official who is responsible for ascertaining first that the weather is suitable, i.e. inclined to rain, and that the wind will not cause the fire to spread to the adjacent forest.

There are, however, occasions when the wind can be of valuable assistance, for instance, in clearing a very damp region. In Brazil, *queimadas* (deliberate forest fires) are started as a preliminary to cultivation. As the forest there is extremely damp, a favourable and rather strong wind (the north-east is the best) is necessary to keep the fire alive. It is the custom to invoke St. Anthony to make it blow in the right direction.

CHAPTER XVII

The Protection of Cultivated Land and Forests

To defend himself from the wind man has no more valuable ally than the vegetable world. In every country in his struggle against it he uses the vegetation that nature has put at his disposal, whether it is for fixing the dunes, sheltering his house or protecting his crops. It is a well-known fact that forests, which influence all the factors of climate, have a particularly marked effect on the wind, keeping it in check and lessening its desiccating properties. From very early times, hedges and rows of trees have been planted to break its force.

Before examining the methods employed for the protection of crops, it is interesting to consider the influence of the wind on vegetation. Agriculture is certainly a sphere of human activity in which the wind plays a most important part. Almost everywhere it has a more or less pronounced influence on crops—sometimes beneficial but more often harmful.

The Wind and Plants

Plants are affected by the wind in many different ways. It reduces their size, modifies their form and structure, facilitates their fertilization and dispersal, regulates their distribution and many even oppose their existence when its force and frequency are too great. The wind is particularly the enemy of trees and is the cause of their total absence on hills beaten by gales and on coasts exposed to storms.

Trees growing at the top of cliffs and subject to the continual action of a wind from a constant direction are strangely deformed and stunted. The curious dwarf forests on the coasts of Newfoundland and of Labrador, where the trees are often not more than a few feet high, are solely due to the action of the wind. Unable to grow in height, the stunted conifers sprawl and trail, intertwining their branches and forming tangled thickets. The absence of trees on certain islands like St. Paul, the Crozet and the Kerguelen in the Indian Ocean and Campbell and Macquarie in the Pacific is also due to the great west winds and not to the latitude, which is comparable to that of the English Channel.

Bent trees may be found not only along the coasts but inland in all very windy districts, in the bottom of valleys, on the slopes of mountains and on very exposed plateaux. It is not necessarily a very strong wind that alters the shape of trees; a wind blowing in an almost constant direction will have the same result.

From the physiological point of view, the wind has a desiccating effect on plants by promoting transpiration and evaporation. The more violent it is, the more they lose their sap. Excessive evaporation may wither them in strong gales and prevents the growth of trees on certain coasts although they are very damp and on some low hills which receive abundant rain.

The drying effect of the wind modifies and sometimes even profoundly changes the nature of plants. In very windy countries, to reduce the surface over which evaporation takes place, they remain small; they are covered with down and thorns and their leaves are diminutive and tough. In the xerophytes of desert areas, which are always subject to dry and violent winds, the leaves become atrophied.

The wind by carrying pollen acts as a fertilizing agent and by carrying seeds over long distances facilitates their dispersal. It may introduce weed seeds into a district where they were formerly unknown and where they increase to such an extent that they become a menace to the crops.

The wind is also injurious to vegetation by drying out, chilling and hardening the soil, by scattering the dead leaves and mould and in extreme cases by removing the loam. Particles of light and sandy soil driven by a violent wind lash and wear away branches and leaves, which may be entirely torn off and destroyed. The wind often buries young plants in the fields. Complaints are made of this kind of damage in the tobacco plantations of Ontario and Quebec, where large areas sometimes have to be replanted, with the result that the crop is late in maturing. In the sandy region of Guardamar in Catalonia vine stocks and fig-trees are frequently partly buried.

During strong gales near the sea, the air is often impregnated with salt and the action of the salt spray carried by the wind scorches the leaves of the trees. The effects of this can be clearly seen on the coasts of tropical seas after the passage of a cyclone. One may form an idea of the distance that the salt can travel by the fact that on 24th November, 1814, the rain falling at Blackwall, which is about 125 miles from the sea, was laden with salt as a result of a storm on the coast. Certain types of grass, however, are unharmed by these particles of salt brought by the sea winds. The salt meadows on the Channel coast of France are renowned for the quality of the mutton that they produce.

A phenomenon of a rather different kind but also attributable to the wind has been observed in the Camargue. The mistral greatly increases evaporation and causes the salt to rise to the surface of the soil, rendering this unfit for cultivation, especially in the southern part of the delta.

Damage Caused by the Wind

Every year and in every country, the wind causes great damage to forests, fruit trees and to cultivations of all kinds. High winds break the branches of the trees and uproot them; they destroy the buds or the blossom of fruit trees or, later, make them drop their fruit prematurely so that part of the crop is lost. To protect their orchards and market gardens from the late frosts brought by the north wind growers in many places instal a heating system at night at the foot of each tree. In other places, the hot winds are the most harmful to crops, like the sirocco in Algeria and Tunisia, or the kamsin in Egypt.

In some years high winds in the spring and autumn cause enormous damage in the forests of Central and Northern Europe.* Among the worst, we may mention the storm of

* *Translator's Note.*—The great gale of 31st January, 1953, destroyed 47,000,000 cubic feet of timber in Scotland (see the Annual Report of the Forestry Commission, 1953).

12th February, 1894, and of 25th December, 1902, which struck down about 14,126,400 cubic feet of timber in Denmark. Those of 1915, 1929, and 1930 destroyed 356,882,635 cubic feet in Czechoslovakia. In Austria, 40,365,000 cubic feet were struck down by hurricanes. In Switzerland the *föhn* brought down, in the canton of St. Gallen alone, 12,360,600 cubic feet between 1919 and 1925. Tropical cyclones also cause terrible damage, especially to coconut palms. In 1926 those in the Philippines were all destroyed by typhoons.

Some other effects of the wind are not so well known. It has been observed in various places that high winds can check the fruiting of trees by preventing the bees and other insects from carrying the pollen from the stamens to the pistils. In Austria, it is thought that by reducing the proportion of carbon dioxide in the atmosphere the wind is causing the edge of the forest to recede in the high mountains.

Windbreaks

Undulations in the terrain, when they are suitably oriented, can provide natural shelters in some particularly windy districts and give valuable protection to the crops. The cultivated areas in many Alpine valleys are situated on the slopes sheltered from the cold north winds. Along the coasts and on many small low-lying islands, like Noirmoutier and Sein, where the wind beats rudely on the vegetation, the gardens lie hidden in the smallest folds in the ground. These privileged positions, however, are relatively limited in number, and the growers often have to make artificial shelters or windbreaks to protect their crops.

The art of struggling against the wind is very old and windbreaks are in use in all countries. There are many varieties of them, but they all belong to two principal types: *living fences* and *dead fences*. The first type is much more common and consists of hedges and rows of trees, formed of the hardiest species which are the most resistant to the wind. The vegetation in every country usually includes a certain number of plants and trees suitable for this purpose and some varieties which are particularly effective as windbreaks are widely used in different countries. The dead fences consist of brick or loose stone walls.

The latter are more common and afford a use for the stones when the ground has been cleared. The numerous walls mentioned by Jean Brunhes that can be seen in Mahon in Minorca are made in this way. All over the island, the stones removed from the ground to make way for cultivated fields and gardens have been used to erect walls from three to six feet high to shelter the plants from the most frequent and formidable north wind.

In many countries, windbreaks consist of hurdles and fences made of stakes, boards, slats or straw matting. In Italy and the Lipari islands the gardens, and particularly the tomato plants, are shielded from the north winds by reeds entwined or simply planted in the ground. In some parts of the United States, especially in Pennsylvania, the tobacco fields and tomato plants are protected by screens made of planks, canvas and glass. Tree trunks or branches placed on earth banks are also used to shelter the gardens. Windbreaks are placed perpendicularly to the direction of the prevailing wind but dangerous winds sometimes come from other directions and it is then advisable to make an interlaced screen. Very often only temporary screens are needed to protect the plants during a period of their growth. Some plants are very sensitive to the wind when they are young but quite capable of withstanding it when they are fully grown.

There are other less known methods of protecting crops. In Holland, for example, the fields of tulips, hyacinths, and narcissi, broken up by protective hedges, are covered with straw in the winter to safeguard them against both the frost and the high winds. In Michigan in the United States, it is customary to sow strips of rye in the midst of fields of onions. Rye is also an excellent windbreak between rows of vines in newly planted vineyards on sandy soil. This sytem has been adopted in Hungary.

Th. Monod describes the curious method of cultivating the vine in funnels, practised in the south of Lanzarote in the Canary Islands. Some districts have a most strange appearance, for the vines, hidden at the bottom of their holes, cannot be seen. The holes are clustered on the slopes of ancient volcanoes and have been dug in the ash to protect the vine from the wind and to reach deeper and more moist layers of soil. In another part of the island, in the neighbourhood of San Bartolome, the vines are protected in a different but equally ingenious fashion against the violence of the trade wind. Each plant is sheltered by a low stone wall in the form of a horseshoe with the convex side facing the wind.

Primitive peoples like the Polynesians on the Wallis and Futuna islands in the South Seas are not unaware of the risks to which their plantations are exposed by high winds. On Wallis Island the plantations are made in the clearings of the forest, which makes an excellent natural shelter. The islanders of Futuna prefer to let their yams trail on the ground rather than use supports because they have observed that the young shoots are then less damaged by the wind. At Shinyanga in Tanganyika, the natives encircle the plantations surrounding their huts with curtains of trees or hedges to break the force of the wind. In some places the cultivated areas are completely enclosed by high hedges, in others there are only parallel lines of trees placed on the side of the prevailing wind.

Screens of Trees

The best method of mitigating the effects of the wind is to plant rows of trees. The trees most commonly used in Europe to form protective hedges and screens against the wind are the willow, the osier, the genista, the hawthorn, the hazel, the poplar, the robinia, the ash, the young elm, the hornbeam, the alder, the birch, the beech, the spruce and the Corsican pine, as well as the yew and the cypress in more southern districts. The maple, the oak, the lilac and the cornel-tree are also sometimes used. The eucalyptus, the cypress, the Aleppo pine and the cluster pine, the mulberry, the pomegranate and the tamarisk are more often found along the Mediterranean coasts. The prickly pear and the agave also make excellent windbreaks, very common in Sicily.

In Egypt, the she-oak, the eucalyptus, the Arabian acacia and the tamarisk as well as beech grass are widely used. Rows of poplars are very common in China. In the United States and Canada, the Siberian acacia and different kinds of conifers (white pine, Banks' pine, Scotch fir, etc.) are used as well as the species most common in Europe.

Screens for protection against the wind are also required in the cultivation of tropical plants. The she-oak (*Casuarina equisetifolia*) is very often found, especially near the sea. At Dahomey, it is arranged in a screen of about twenty rows along the edge of the coast to protect the coconut palms and enable them to grow more quickly. It is used for the same purpose in the Philippines. On Reunion, rows of she-oaks protect the sugar-cane plantations. On the Ivory Coast, mango trees and lemon trees give good results as screens. In French Guinea, where the *harmattan* causes great damage to pineapple and coffee plantations, screens of banana trees are used, while in the neighbourhood of Dakar the banana trees are shielded by she-oaks. Certain spurges make screens which effectively protect the kitchen gardens in French West Africa. The *Cassia siamea*, which has long lateral branches from the base of the trunk and remains green throughout the dry season, makes the best windbreak on the high plateaux of the Belgian Congo. Planted in rows about thirteen feet apart between the fruit trees, it grows quickly to form thick hedges which keep back the wind perfectly.

On Madagascar, good results are obtained by planting rows of mango trees, bread-fruit trees and acacias. In the Philippines, besides the she-oak already mentioned, the bamboo, also found in many other tropical countries, the croton and different local plants are used to form hedges. The banana serves to protect the coffee and cocoa plantations. In Australia, the Aleppo pine, various acacias, the eucalyptus and the camphor tree are used to make windbreaks; the almond tree and the olive tree are preferred in orchards.

Advantages and Disadvantages of Windbreaks

There used to be a complaint against windbreaks that they restricted the area of cultivatable land, but the space taken from crop growing by hedges or screens of trees is largely compensated by the wood that is derived from them. Some people consider them harmful because they shade the crops and because their roots spread and take too much nourishment from the soil. It is easy to remedy this disadvantage by providing extra manure or even by cutting the roots on the side of the crops. In Belgium, during hard winters, the screens of trees are said to prevent the wind from shaking off the snow which accumulates on the top of the young pines with the result that they are broken by its weight.

On the whole, however, the disadvantages of windbreaks are comparatively slight.

The observations made in various countries concerning the effects of the wind on cultivations, especially in Denmark and Holland, where a special interest is taken in the subject, have proved that screens of trees afford the best means of protection. Their useful effect extends over a variable width, generally estimated as between six and ten times their height. It is generally agreed that in Luxembourg the adjacent territory is screened over a distance equivalent to only three times the height of a windbreak, in Switzerland four times and in Canada five times. This distance increases to eleven times in Denmark, to ten or twenty times in Hungary and to nearly thirty times in Tripolitania. It has even been claimed that in Senegal hedges of she-oaks protect the ground over a distance of fifty or a hundred times their height. These very different estimates show that the efficacy of windbreaks depends on the topographical conditions and on the species of trees of which they are formed.

Windbreaks have many different advantages. In addition to breaking the force of the wind, they increase the temperature and humidity of the air and of the soil and reduce evaporation. The yield from orchards and market gardens is generally higher when they are sheltered. It was noticed in Hungary that the crops were earlier when protective screens were used. In French Guinea, the banana groves shielded from the east wind double their yield. The cultivation of fruit trees is said to be almost impossible in Senegal without windbreaks. In the Province of Quebec, the use of windbreaks is very effective in the cultivation of tobacco; they enable the plant to develop more quickly in the summer and to keep its leaf intact until harvest time. In Canada, the snow generally accumulates on the screens of trees and when it melts distributes a quantity of water which makes the soil more moist in the summer.

Besides being useful in agriculture, windbreaks have a considerable influence on the health of animals and when they are used the flocks are better nourished. This is not a recent observation; in England more than a hundred years ago the rent of pasture land was higher when it was surrounded with hedges. In Uruguay, the stock-rearers are careful to ensure the protection of their animals by screens of trees. The New Zealand farmers, who leave their herds outside all the winter, take the precaution of surrounding the pastures with hedges so that the sheep and cows can find shelter whatever the direction of the wind may be. The cows are also covered with sacking to protect them from the rain as long as the bad weather lasts. In Patagonia, the sheep are also left out all the winter, but special shelters from the wind are arranged for them. In some countries, too, windbreaks are widely used for apiaries.

In addition to these advantages connected with the climate there are some economic advantages. Firewood and wood for carpentry, posts for fences and, in some cases, fruit may be obtained from the trees. The eucalyptus, used as a windbreak in countries often very poorly wooded, gives a wood very highly esteemed. Similarly, the she-oak, which, on Reunion, shields the fields of sugar-cane, is used as a fuel in the sugar factories. In the west of Canada and the United States, the presence of protective screens greatly increases the value of the farms. At Malacca, windbreaks are considered to localize the damage caused by insects and cryptogamic diseases to the rubber plantations.

The Protection of Forests

The damage caused by the wind is not confined to crops but also affects the forests which have to be safeguarded. Old plantations of trees have a particularly poor resistance to the wind. The ravages caused by storms, in addition to the loss of wood, involve the cost of restoration and favour the invasion of harmful insects.

One of the methods recommended to reduce the risk of damage by the wind is to make mixed plantations of trees consisting as much as possible of tap-rooted varieties. In recent years gales in Europe have caused widespread destruction in many forests of single varieties, especially fir trees. Leafy trees are more resistant than conifers and mixed forests of beech and oak are among those least affected.

Forests are also shielded from violent winds by surrounding them with tree screens and windbreaks, planted perpendicularly to the prevailing wind and consisting of the same varieties as those used to protect crops. In Belgium, for example, the young plantations of Scotch firs are protected by screens of spruce, planted at intervals of three to three and a half feet in one, two, three, or five rows. In several countries a special arrangement is made on the edge exposed to the dangerous wind. One method used in Pennsylvania is to plant trees such as maple, elm, catalpa, spruce, black pine, etc., in four rows and allow their low branches to trail on the ground. Very often the felling of trees on the edge of the forest is forbidden so that they may be left to form a protective screen. Similarly in Ireland, the trees are planted very close together on the edge of the forest over a width of 165 to 330 feet; sometimes on a narrower belt sixteen to thirty-five feet wide they are polled so that their tops and branches, being more developed, offer an obstacle to the wind. In other places, a clearing is made near the edge, leaving the trees with many branches growing from the base to develop more vigorously and provide a screen.

In narrow valleys, cutting should be done against the prevailing winds in a line following the direction of the contour lines. Generally speaking, it is always an advantage to fell the trees in bands, keeping the edge intact as long as possible on the side from which the strongest wind blows.

The science of defence against the wind has become more important in recent years and all means are being tried to reduce the damage caused to trees in storms.

CHAPTER XVIII

Mythology and Legend

THE wind has always struck forcibly the imagination of men and the farther we go back in history to the religious conceptions animating the forces of nature, the greater the role of the wind appears to have been, as it still is today in the beliefs of many primitive peoples.

The important role of the winds in agricultural and maritime life caused them in very early times to be personified in Greek and Roman mythology. Mythologists agree that at the beginning Hermes and other powerful gods, in which the great forces of nature were believed to be incarnated, were representations of the wind. Only later did the winds become special deities subject to the great gods. They were Boreas, Notus, Eurus and Zephyrus, beneficent winds, the sons of Astraeus and Aurora (dawn). These were all that were known in the Homeric era and were placed under the sovereignty of Jupiter and under the direct orders of Aeolus, the god of the sea winds. To them must be added the baleful winds, represented by Typhoeus or Typhon, the god of the storm. All these winds were the object of formal cults in different places. They were localized in mountainous regions in relation to the direction from which they blew over Greece. The most violent dwelt in the mountains and gorges of Thrace.

In the second century B.C. there were eight winds. They were outlined in bold strokes on the Clock of Andrinokos Chyristes at Athens, better known as the "Tower of the Winds". Draped in flowing robes and gliding with wings on their shoulders in the directions which characterized them were:

Boreas (N.), Kaikias (N.-E.), Apheliotes (E.), Eurus (S.-E.), Notus (S.), Lips (S.-W.), Zephyrus (W.), and Skiron (N.-W.). The symbolical attributes of the figures on this tower correspond to the most typical meteorological characteristics of each wind and Victor Hugo was able to write: "In the bas-reliefs of the Tower of the Winds, the icy winds are hideous and shaggy; they have a stupid air and are clothed like barbarians; the soft warm winds are dressed like Greek philosophers."

Victor Bérard has shown the large place taken by the winds in the works of Homer and points out that most of the passages having a meteorological character are concerned with winds and tempests. The few hundred verses that Homer devotes to them were for nearly 3,000 years all the meteorological science that the majority of writers had. J. Rouch observed the curious coincidence between Homer's Odyssey and the Nautical Instructions for the Greek seas, in which is found the justification for the long voyage of Ulysses with a favourable wind in the summer.

Greek mythology attributed to Aeolus, as we have seen, the power of commanding the winds, which he kept imprisoned in the caves of his island, unleashing or calming them at his pleasure. The island of Aeolus was later identified as being one of the Lipari Islands, the Aeolian Islands of antiquity, where Vergil shows us the god holding the winds captive at the bottom of a cave (Aeneid, I, 52 *et seq.*). The famous voyager, Ulysses, as recounted in Homer's Odyssey (Book X, *et seq.*) came to consult Aeolus before undertaking his first voyage. Aeolus, whose favour Ulysses had known how to win, gave him a bag containing the winds, cautioning him not to open it without pronouncing certain words which would release only the winds he desired. The companions of Ulysses, moved by curiosity, took the opportunity when Ulysses was asleep, to half open the famous bag from which all the winds escaped, unleashing continuous storms which helped to delay the return of Ulysses for ten years. A similar legend existed among sailors in the East.

The Romans always had a cult for the winds and dedicated temples to the *Tempestates* (Storms). In literary times, the names of the gods, borrowed from Greek mythology, were substituted for the naïve personifications of primitive periods.

N

Some of the winds, however, kept their old names, like Aquilo (N.-E.), Favonius (W.), and Auster or Africus (S.).

In Slav mythology, Stribog was the god of the winds. Hindu mythology had several, who were named Rudra, Vogu, and Vata.

P.-E. Victor (Ref. 71) says that the Eskimos of Angmassalik, on the east coast of Greenland, know four principal winds, which they name and personify in the following way: Nekrayak (N.-E.) is male, Pittarak (N.-W.) is his wife, Kadannek (S.-E.) is their son and Pouwanguartek (S.-W.) their daughter. When the wind named Kadannek begins to blow, it is said that he is going to meet his mother. In former times, when the Nekrayak lasted too long, an old woman would sleep out of doors naked to apease him; a man would do the same if it was desired to calm the Pittarak.

Certain Australian tribes sometimes considered natural phenomena, particularly the wind, as totems. In North America, the totem of the Creeks Indian tribes is also the wind. The Aïnos, who have a cult for all the great natural forces, worship, among other phenomena, the typhoons which periodically strike the coasts of their archipelago, causing considerable damage.

It is not only for reasons of comfort that certain peoples, the Chinese and the Annamites in particular, orient their houses so that the entrance is shielded from certain winds; it is because they believe these winds to be endowed with supernatural powers.

Volumes could certainly be written on all the legends to which the wind has given rise, without mentioning the innumerable proverbs. Every country has its own, showing the importance attached to this phenomenon. Its power has always seemed striking to man; writers have given it a large place in literature and musicians like Rossini and Mendelssohn have imitated its tumult and wailings.

To close this chapter I will quote two legends, taken from very different sources but both having a certain picturesque quality. The first concerns King Solomon, whose power made a vivid impression on the oriental imagination. When he ascended the throne of Israel, eight angels came to meet him in a valley between Hebron and Jerusalem and assured him of their devotion. They were the angels controlling the winds and they made him the gift of a stone on which were engraved these words: "When you need us, turn this stone towards the sky and we will run to your aid." With his power of commanding the winds, he could travel through the air on a flying carpet, where he took his place with all his court. It was in this way that he went to visit Damas.

The second legend, reported by L. Lacroix, was told by old sailors fifty years ago. In the first centuries of the world, they said, the sea was always calm. A very light wind, similar to the land breeze which blows at night in fine weather, served to take ships in one direction, while the sea breeze during the day took those which were going in the other. Once upon a time, a pirate captain, in pursuit of a ship, sold his soul to the devil, in exchange for winds of his own choice which would enable him to capture his prey without the risk of seeing it disappear in the night. He could thus shut up in his hold all sorts of winds which he used at will, but the prisoners soon became rebellious. One day the pirate ship was sunk and its captain hanged, but the winds, having escaped, began to agitate the sea, taking pleasure in causing suffering to ships and crews. The only remedy, said the legend, was to pray to the All-Powerful, who was stronger than Satan. Touched by the sailors' supplications, God accorded to St. Peter, who was an old sailor himself, the power of making himself obeyed by the unleashed winds. Subsequently certain saints intervened directly in favour of their protégés. This is the origin of the prayers to St. Anne, to Our Lady, to St. Clement, to St. Anthony and to so many others.

Conclusion

SOMEWHERE I read that "the wind is the friend of man." Let us agree that it is a friend to be feared whose benefits are dearly bought. Many people, however, see in this natural force only a cheap and inexhaustible source of energy, which, captured by man, has for centuries driven his ships and turned his mills. These are certainly incontestable services, but the greatest benefit that humanity owes to the wind is of another kind. It is the wind which circulates over the land the immense masses of water vapour evaporated from the sea, bringing us the rain and moisture on which largely depends the distribution of the population over the earth's surface.

In comparison with the two great uses of the wind in the past, the services which we ask of it today are very small. To help in the irrigation of certain arid countries and in the drainage of the Dutch polders, to provide light and water on a few farms in the country—these are at present the principal applications of the wind. It is little enough for a force which is practically inexhaustible.

For a long time men had only an empirical knowledge of the wind, once considered a pre-eminently capricious and inconstant element. The progress of meteorology has shown us that it obeys well-defined laws. This recently acquired scientific knowledge has considerable practical importance. The services that it has rendered in the sphere of aviation are well known and we are now better equipped to protect ourselves against the dangers to which we are exposed by the violence of the wind. Who knows whether, perhaps in the near future, we shall not be able to return to the use of the wind, almost completely neglected in our age, and to exploit its motive power in a more rational and effective manner than in the past?

Not to use the wind is not the same as ignoring it, and we have taken defensive measures to forestall or lessen its effects. They are felt everywhere, with terrible consequences in some regions. All things considered, we will conclude by saying that the wind appears sometimes as a friend of man, sometimes as an enemy, but more often in the latter role.

196

Bibliography

1. ANONYME. — Protection des forêts et des cultures contre le vent. 264 pp., Rome, 1933.
2. AIMES (A.). — Météoropathologie. Maloine, Paris, 1932.
3. AUFRÈRE (L.). — Les dunes sahariennes et les vents alizés. *Bull. Ass. Fr. Avanc. Sc.*, n⁰ 112, pp. 131-8, mai 1933.
4. BEAURIENNE (A.). — La ventilation et le refraîchissement des locaux habités dans les pays chauds. *Bull. Soc. Ing. Civ. Fr.*, n⁰ˢ 11-12, pp. 1727-45, 1931.
5. BÉNÉVENT (E.). — Bora et Mistral. *Ann. de Géographie*, n⁰ 219, t. XXXIX, pp. 286-98.
6. — Le climat des Alpes Françaises. *Mémorial de l'O. N. M.*, n⁰ 14, 1926.
7. BESSON (L.). — Le climat du littoral français de la Manche. *Inst. d'Hydr. et de Climatologie*, n⁰ 3, pp. 211-28. juillet-décembre 1929.
8. — L'altération du climat d'une grande ville. *Ann. Hyg. Publ. Industr. et Sociale*, août 1931.
9. BOUGHNER (C.-C.). — The climate of Canada. *Quart. Journ. of the R. Meteorol. Soc.*, vol. LXIII, n⁰ 271, pp. 419-31, 1937.
10. BRAZIER (C.-E.). — Quelques données climatiques sur la ville de Paris à l'usage des urbanistes. *Rev. Géogr. Phys. et Géol. Dyn.*, vol. IX, fasc. 3, pp. 333-46, 1936.
11. BRUNHES (Jean). — A Majorque et à Minorque. *Rev. des Deux-Mondes*, 1ᵉʳ nov. 1911.
12. — La Géographie Humaine. F. Alcan, Paris, 1934.
13. BUFFAULT (P.). — Les dunes de la Nouvelle-Zélande. *Rev. des Eaux et Forêts*, janvier 1932.
14. BUFFON. — Histoire Naturelle. Nouv édit., t. II, Imprimerie Royale, Paris, 1769.
15. BURGAUD (M.). et GHERZI (B.). — L'annonce des typhons dans les mers de Chine par l'Observatoire de Zi-Ka-Weï. *Journ. Marine Marchande*, 13-27 août 1931.
16. BURON (E.). — Les navigations du Roi Salomon. *Terre, Air, Mer*, t. LIX, n⁰ 3, mars 1933.

17. CATTALA (L.). — Que sait-on des efforts du vent sur les bâtiments et les ouvrages d'art? *Rev. Quest. Scientif.*, pp. 382–423, 20 mai 1937.
18. CHAMPLY (R.). — Les moteurs à vent. Dunod, Paris, 1933.
19. CHAPPUIS (J.).—Ventilation des espaces souterrains. *Journ. des Usines à Gaz*, 20 juin 1933, pp. 265–70.
20. CHAVÉRIAT (R.).—La voile. Grasset, Paris, 1937.
21. CHASE (S.).—Disaster rides the Plains. *The Americ. Magazine,* Sept., 1937.
22. CHILCOTT (E.-F.).—Preventing soil blowing on the southern Great Plains. *Farmers Bull*, nᵒ 1771, United States Dep. of Agriculture, Washington, March 1937.
23. COMBIER (Ch.).—La climatologie de la Syrie et du Liban. *Rev. Géorg. Phys. et de Géol. Dyn.*, vol. VI, fasc. IV, pp. 319–46.
24. CONSTANTIN.—Le vent, source inépuisable d'énergie à bon marché. *La Nature*, pp. 395–400, 21 juin 1924.
25. COURVILLE (R.).—Le Moulin de la Galette va entrer dans l'histoire de Paris. — *Intransigeant*, 31 mars 1939.
26. DEFFONTAINES (P.), JEAN BRUNHES-DELAMARRE (M.), BERTOQUI (P.).—Problèmes de Géographie Humaine. Bloud et Gay, Paris, 1939.
27. DEMANGEON (A.).—L'habitation rurale en France (Essai de classification des principaux types). *Ann. de Géographie*, pp. 352–75 1920.
28. DEVAUX (P.).—«Astuces» techniques et sportives dans la navigation à voile. *Sciences et Voyages*, nᵒ 26, pp. 85–8, août 1937.
29. DION (R.).—Le Val de Loire. 752 pp., Arrault et Cⁱᵉ. Tours, 1934.
30. FOVILLE (A. de).—Introduction de l' «Enquête sur les conditions de l'habitat en France. (Les maisons types)». Ministère de l'Instruction Publique. 381 pp. E. Leroux, Paris, 1894.
31. FROIDEVAUX (H.).—Les étapes de l'utilisation des moussons. *La Géographie*, mai 1932, t. LI, nᵒˢ 5–6.
32. FRUH (J.).—Geographie der Schweiz (Natur des Landes). St. Gall. 1930.
33. GAUTIER (E.-F.).—Le Sahara. Payot, Paris, 1923.
34. — L'Afrique Blanche. A. Fayard, Paris, 1939.
35. HOPKINS (E.-S.) and BARNES (S.).—The rotation of crops and the cultivation of the soil in the Prairie Provinces. Ottawa, Ministry of Agriculture. Bull. 98, New Series, 1928.
36. HOULLEVIGUE (L.).—Les principes de la météorologie. *Rev. de Paris*, 1ᵉʳ mars 1908.

37. Hugo (V.).—L'homme devient oiseau. *Rev. de Paris*, 15 avril 1910.

38. Joel (A.-H.).—Soil conservation reconnaissance survey of the Southern Great Plains wind erosion area. *Techn. Bull.*, nº 556, January, 1937, U.S. Department of Agriculture, Washington.

39. Julien (G.).—L'habitat indigène dans les possessions extérieures de la France. Extrait de *La Terre et la Vie*. Soc. Edit. Géogr. Maritimes et Coloniales, Paris, 1931.

40. Jurien de la Gravière (E.).—La navigation hauturière. *Rev. des Deux-Mondes*, sept. 1874.

41. — La marine de l'avenir et la marine des anciens. *Rev. des Deux-Mondes*, 1ᵉʳ août 1878.

42. Kronfeld (R.).—Le vol à voile. Gauthier-Villars, Paris, 1935.

43. Lacroix (G.).—Les moteurs à vent. *La Nature*, nº 2823, 15 déc. 1929.

44. Lacroix (L.).—Les derniers grands voiliers (Histoire des Long-Courriers nantais de 1893 à 1931). J. Peyronnet et Cⁱᵉ, Paris, 1937.

45. La Roërie (G.).—A l'aube des grandes découvertes. *Terre, Air, Mer*, avril 1932.

46. Leenhardt.—Gens de la Grande Terre. N. R. F., 1937.

47. Malycheff (V.).—Le loess. *Rev. de Géogr. Phys. et de Géol. Dyn.*, vol. II, fasc. 2, juin 1929; vol. III, fasc. 4, 1930; vol. IV, fasc. 3, 1931; vol. VI, fasc. 2, 1933.

48. Marchand (E.).—Les vents dans le Sud-Ouest et le Sud de la France. Le Vent d'Autan. *Bull. Obs. Carlier d'Orthez*, 1902–1903.

49. Marcy (G.).—Notes linquistiques relatives à la terminologie marocaine indigène des vents. *Mém. Soc. Sc. Nat. Maroc*, nº XLI, 15 sept. 1935.

50. Marié-Davy (H.).—Les mouvements de l'atmosphère. Masson, Paris, 1877.

51. Martonne (E. de).—Traité de Géographie Physique. A. Colin, Paris, 1923.

52. — Essai de classification des grands types de climats. *Sciences*, nov. 1937.

53. Missenhard (A.).—L'Homme et le Climat. Plon, Paris, 1937.

54. Monod (Th.).—Notes Canariennes. *La Terre et la Vie*, nº 8, pp. 451–68, août 1934.

55. Mouriquand (G.).—Les variations atmosphériques et leur rôle biologique. *Sciences*, 1937.

56. MOUSSET (A.).—Les moulins à vent parisiens. *Journal des Débats*, 8 avril 1939.
57. NORMAND.—The effect of high temperature, humidity and wind on the human body. *Quart. Journ. of the Royal Meteorological Soc.*, January, 1920.
58. PASTEUR-VALLERY-RADOT.—Météorologie et médicine. *Rev. des Deux Mondes*, mars 1937.
59. PAULY (P.-Ch.).—Climats et endémies. Esquisses de climatologie comparée. Masson, Paris, 1874.
60. PETITJEAN (L.).—Vents de sable et pluies de boue. *Sciences*, mars 1937.
61. RABOT (Ch.).—La tempête du 31 décembre 1904 dans la Baltique occidentale. *La Géographie*, t. XI, pp. 32–8, 1905.
62. RADAU (R.).—Le rôle des vents dans les climats chauds. Gauthier-Villars, Paris, 1880.
63. RAMBERT (E.).—La question du foehn. *Biblioth. Universelle*, no 127, 1er juillet 1868.
64. RECLUS (E.).—Le littoral de la France. *Rev. des Deux-Mondes*, 15 déc. 1862 et 1er août 1863.
65. RENOUARD (J.).—Le Mistral. *Journ. des Débats*, 22 février 1939.
66. RICKARD (Th.).—L'Homme et les Métaux. N. R. F., 1938.
67. SANVOISIN (G.).—Les collectionneurs de girouettes. *Journ. des Débats*, 10 octobre 1938.
68. SAVIGNON (L.).—Les phénomènes météorologiques en pathologie humaine. Paris, 1935.
69. SCAËTTA (H.).—La genèse climatique des sols montagnards de l'Afrique centrale. *Inst. Royal Colonial Belge* (Sect. des Sc. Nat. et Méd. Mémoires), t. V, Bruxelles, 1937.
70. THOENE (P.).—Les hommes volants. *Les Annales*, no 2610, p. 272, 1938.
71. VICTOR (P.-E.).—Boréal. Grasset, Paris, 1938.
72. VIDAL DE LA BLACHE (P.).—Principes de Géographie Humaine. A. Colin, Paris, 1922.
73. WARMINGTON (E.-H.).—The commerce between the Roman Empire and India. Cambridge, University Press, 1928.
74. WEBSTER (H.).—La grande détresse de nos moulins. *L'Illustration*, 2 oct. 1937.

Index